Digital Cinema

Digital Cinema

The Revolution in Cinematography, Postproduction, and Distribution

Brian McKernan

McGraw-Hill

New York   Chicago   San Francisco   Lisbon   London   Madrid
Mexico City   Milan   New Delhi   San Juan   Seoul
Singapore   Sydney   Toronto

The McGraw-Hill Companies

Cataloging-in-Publication Data is on file with the Library of Congress.

Copyright © 2005 by The McGraw-Hill Companies, Inc. All rights reserved. Printed in the United States of America. Except as permitted under the United States Copyright Act of 1976, no part of this publication may be reproduced or distributed in any form or by any means, or stored in a data base or retrieval system, without the prior written permission of the publisher.

2 3 4 5 6 7 8 9 0   IBT/IBT   0 1 0 9 8 7

ISBN 0-07-142963-8

The sponsoring editor for this book was Stephen S. Chapman and the production supervisor was Sherri Souffrance. It was set in Helvetica by Patricia Wallenburg. The art director for the cover was Anthony Landi.

Printed and bound by IBT Global.

 This book was printed on recycled, acid-free paper containing a minimum of 50% recycled, de-inked fiber.

McGraw-Hill books are available at special quantity discounts to use as premiums and sales promotions, or for use in corporate training programs. For more information, please write to the Director of Special Sales, McGraw-Hill Professional, Two Penn Plaza, New York, NY 10121-2298. Or contact your local bookstore.

To Elizabeth, my favorite writer

Contents

Preface

Humans are a storytelling species. Throughout human history, stories have served to sustain cultures and traditions and to explain the mysteries of life and the world in which we live. And from *Gilgamesh* to the latest TV commercials, whether in the mind's eye or on a screen, stories either evoke or are conveyed using images.

During the past century, moving-image media greatly expanded the ways in which stories could be told. The huge impact of film and television in commerce and culture is undeniable. Now, today's digital technology has democratized this most powerful form of storytelling, making it affordable enough for practically anyone to use.

"With digital it's really been giving a democracy to filmmaking," observed Spike Lee, famed director of such films as *Do The Right Thing* and *Bamboozled*, at a recent media conference. "It's not…to say we're going to have better films, but more people will be able to make films. And so to say you don't have enough money to make a film is no longer an excuse. If you have an idea, you can make a film."

This book provides a look at today's digital cinema revolution, the technology and acceptance of which is advancing with ever-greater speed. There is no single definition of *digital cinema*; it generally tends to involve four major categories, which are the digital photography, post-production, distribution, and exhibition of theatrical movies. Although film may or may not be involved in any or all of those categories, film is very much a part of the digital cinema revolution.

Digital cinema is a term that can also apply to Hollywood movies just as much as it does to indie filmmaking, and it can include everything from relatively inexpensive productions shot with "prosumer" camcorders to big-budget effects-laden movies made with the most advanced technology. What all have in common is the flexibility of digital imaging to render onscreen whatever the mind can conceive—and, if desired, to do so affordably.

I hope that the contents of this book will inspire you to pursue your creative vision and use the tools of digital cinema to tell a story that will help to make the world a better place.

Acknowledgments

Special thanks to: George Avgerakis for helping to get *Digital Cinema* started, Scott and Minky Billups for their incredible support when I needed it most, Steve Chapman for his patience, Dan DiPaola for being a digital pioneer, Mark Forman for his superb photography and generosity, Paul Gallo and Martin Porter for believing in the digital cinema concept, Robert Goodman for his camera expertise, Debra Kaufman for her postproduction knowledge, Dale Launer for his insights, David Leathers for his intuition and reliability, Jim and Charlene Mathers for their kindred spirit of enthusiasm for digital cinema and their tireless efforts on behalf of the Digital Cinema Society, Dick Millais for his cheerful encouragement, Pete Putman for his mastery of projection technology, John Rice for providing me with extraordinary opportunities, Toby Sali for his unwavering support, Mark Schubin for all that he has taught me, Peter Symes for sharing his brilliance, Tim Tully for his audio advice, Bob Turner for his unsurpassed knowledge of editing technologies, and Bob Zahn for his selfless cooperation and his dedication to the art of the moving image. Most of all, thanks to Megan for everything.

Portions of this book were adapted and updated from articles I originally wrote for *Digital Cinema* magazine.

Beginnings

Humans have been making images for at least 30,000 years. The delicate and colorful Ice Age cave paintings at Lascaux, France and Altamira, Spain attest to the innate human tendency for image making, which has played a vital role in the evolution and development of civilization (Figure 1-1).

Pictographic communication eventually gave rise to the alphabets, languages, and arithmetic that launched the world's cultures. Advances in image-making were integral to civilization's growth, with carved and sculpted stone, cast metal, paint, mosaics, and other materials providing a means by which record-keeping, ritual, religion, art, and other human endeavors found expression. Confucius is alleged to have noted more than two millennia ago that a single picture can be worth 100,000 words. The intellectual and scientific progress of every culture has been reflected in the images it creates, and these pictures can continue to communicate information millennia after the civilization itself disappears.

Ancient Wisdom

During the late 13th century, the art of classical Greece and Rome inspired the beginnings of Europe's Renaissance. It was a time when trade advanced sufficiently to enable a merchant class to support the work of artists. Inspired by the surviving statues and imagery of the great civilization that had flourished in and around their homeland during the Classical era, Italian painters and sculptors began to employ perspective and a realistic style that contributed to a new appreciation for the natural world. And what began as new ideas about viewing man and nature eventually encouraged advances beyond art into the

sciences, commerce, philosophy, and many other fields. This rebirth in attitudes brought forth the Reformation and—eventually—our own modern age.

Renaissance image making also advanced the foundation for a technology that would someday produce moving images: the *camera obscura*. Literally defined as "dark chamber," the earliest surviving evidence of its existence comes from the writings of the great 10th century Persian mathematician Alhazen (Abu Ali Hasan Ibn al-Haitham), who is also credited with originating the science of light and lenses.

Simply stated, a camera obscura is a box with a pinhole in one side through which light enters and reflects on the opposite interior wall; in so doing, it projects an image within the box of whatever objects are external to the pinhole. Leonardo da Vinci and other Renaissance artists used camera obscuras to trace real-world scenes before committing them to canvas. In 1589, the Neapolitan scientist Giovanni Battista della Porta, in the second edition of his *Magiae Naturalis*, may well be predicting movie theaters; he notes that within a large enough camera obscura, an entire audience "may see hunting, battles of enemies, and other delusions."

A century later, German Jesuit Athanasius Kircher described a lantern-illuminated "slide projector" (the *magia cystera*), and by the late 17th century itinerant projectionists carried portable magic lanterns in backpacks from town to town to project pictures painted on glass slides for paying audiences. Magicians, skilled for millennia in the art of illusion, were among the first to inventively exploit what could be done with magic lanterns. In 1790, Belgium's Etienne Gaspard Robert (known as "Robertson") terrified spectators when he used magic lanterns known as *Fantasmagories* to project ghostly images of heroes of the French Revolution onto gauze and clouds of smoke in a Paris cemetery. French magician Jean Robert-Houdin founded a theater for this purpose in Paris in 1845, which helped prepare audiences for the coming of motion pictures decades later.

Early Discoveries

As the Industrial Age dawned, improvements in printing and early photographic techniques made the creation and dissemination of images more accessible and widespread. Whether rudimentary motion imaging was understood in antiquity is currently unknown. We do know, however, that in 1824, British doctor Peter Mark Roget (creator of *Roget's Thesaurus*) published a document titled "The Persistence of Vision With Regard to Moving Objects." We also know that fellow physician Dr. John Ayrton Paris popularized the thaumatrope ("turning marvel"), a simple disc-shaped toy that demonstrated Roget's principle. Although such toys probably existed long before either man was born, Paris showed that when spun, separate images on either side of the disc appeared to be in the same picture.

Two years later, French scientist Joseph Nicéphore Niépce created history's first photograph, using a camera obscura to capture the view from his workroom window on a light-sensitive pewter plate. He soon partnered with fellow Frenchman Louis Daguerre, and through trial and error, they and others gradually improved photographic materials and techniques.

The thaumatrope did not depict motion, but as the 19th century progressed, devices that did proliferated, products of the Age of Invention that was also yielding such wonders as the steam engine, railroad, and telegraph. Early motion-image devices included Belgian professor Joseph Plateau's Phenakistiscope (1832), British educator William George Horner's Daedatelum (1834), and French inventors Pierre Desvignes'

Zoetrope (1867) and Emile Reynaud's Praxinoscope (1877) (Figure 1-2).
The Phenakistiscope used a rotating disc, the Zoetrope, Daedatelum, and
Praxinoscope rotating drums; all displayed sequential images that were
viewed through slots. In the United States, Coleman Sellers' 1861
Kinematoscope employed a paddle wheel. Magic lantern pioneer L.S.
Beale's Choreutoscope (1866) used a strip of sequential slides and a
shutter, presaging the motion-picture projector by nearly 30 years. In
1870, Henry R. Heyl invented the Phantasmatrope, a magic lantern device
that projected what may have been the first-ever photographic motion pictures onto a screen at Philadelphia's Academy of Music for an audience
of more than 1,500. Emile Reynaud combined his Praxinoscope with a
magic lantern in 1892 to create an early movie theater in Paris.

Dawn of an Industry

Persistence of vision is generally believed to be an optical illusion by
which the human eye and brain interpret rapid sequential still images as

continuous motion. It is what makes possible our perception of motion pictures and television, the images of which are comprised of many individual "frames" of projected film or scanned pictures created within a cathode-ray tube or other video display device. (Debate lingers as to whether persistence of vision is physiological or psychological.)

While motion-image display devices were rapidly evolving during the 19th century, so too were advancements in photography. It had become a profession by 1872, when Eadweard Muybridge, a San Francisco-based practitioner of that art, was hired by California millionaire Leland Stanford to settle a bet. Stanford (founder of the university that bears his name) had wagered a friend that at certain points in time, galloping horses raise all four hooves off the ground simultaneously. Muybridge set up a series of cameras along a track, triggering shutters at regular intervals and obtaining photographic evidence to prove Stanford correct (Figure 1-3). The photographer later published many other animal- and human-locomotion studies and created what he called the Zoopraxiscope to project these images.

Muybridge's work attracted the attention of Thomas Edison, 10 years after his 1877 invention of the phonograph. Having won tremendous acclaim for that marvel, Edison was inspired to begin work on creating a

Figure 1-3

Eadweard Muybridge used sequential photography in 1872 to prove that galloping horses raise all four hooves off the ground simultaneously. (© *Corbis*.)

device that could capture motion pictures and also record synchronous sound. Edison later wrote that "it was possible to devise an instrument which should do for the eye what the phonograph does for the ear, and that by a combination of the two all motion and sound could be recorded and reproduced simultaneously."

Edison assigned the task to William Kennedy Laurie Dickson, a French-born Scot who had immigrated to America specifically to work for the great inventor. Dickson experimented with ways to make a camera and a viewing device. He convinced the Rev. Hannibal Goodwin, an Episcopalian minister and chemist, to share his invention of flexible transparent celluloid (popular for use in shirt collars) coated with photographic emulsion with George Eastman, founder of Kodak. It is popularly believed that Eastman asked Edison how wide he should make these flexible film strips and the inventor responded by indicating the desired width with his thumb and forefinger. Eastman interpreted the distance to be 35mm, which today continues to be the leading worldwide format for motion-picture film.

Dickson built his first camera—the Kinetograph—in 1890 and made a five-second silent film of coworker Fred Ott sneezing. Next came the Kinetoscope (Figure 1-4), a peep-hole viewing device that took advantage of another Edison innovation—the electric light bulb—for illumination. A modification, the Kineto-Phonograph, offered sound reproduction, and

Figure 1-4

Kinetoscope arcades were instrumental in introducing the moving image to mass audiences.
(© *Corbis.*)

Dickson set about photographing his coworkers while also recording the sounds they made on a wax cylinder. One such film survives; the 17-second *Dickson Experimental Sound Film* (1894) shows one Edison employee playing the violin while two others waltz. (Oscar-winning film and sound editor Walter Murch transferred and resynchronized the original film strip and sound cylinder to videotape in 2000, reuniting these elements for the first time in over a century.)

Dickson also built Edison's "Black Maria," the first motion-picture studio, in West Orange, NJ. This ungainly black tar-paper shed was designed to rotate so it could stay aimed at the sun for illumination throughout the day. By 1894, Edison's company was busy shooting film shorts and building Kinetoscope viewers for the many penny arcades and peep shows opening to cash in on this new and novel diversion. Dickson's relationship with Edison soured and he formed his own company, American Biograph, which manufactured the competing Mutoscope viewer, a hand-cranked flip-book device.

In France, meanwhile, brothers Auguste and Louis Lumière built a combination camera-projector called the *Cinematographe* (derived partly from the Greek word for movement; it would later be shortened to *cinema* as an all-encompassing term for the art and technology of filmmaking and exhibition) and the first mass-produced movie cameras. They also made their own films, causing a sensation when they premiered them in Paris on December 28, 1895. Even though these first films—with titles such as *Leaving the Lumiere Factory* and *Arrival of a Train at the Station of La Ciotat*—showed such mundane scenes as workers leaving their father's automobile factory at the end of the day and a train stopping at a railroad station, audiences were astonished. On April 23, 1896, Edison's Kinetoscope projector was unveiled at Koster & Bial's Music Hall, in New York City. The following day's edition of *The New York Times* described the Kinetoscope's images as "wonderfully real and singularly exhilarating."

Other inventors and showmen, meanwhile, were making and presenting movies in London, Berlin, Brussels, and elsewhere. The moving image was catching on fast in an age when people were excited by other new inventions such as the telephone, the player piano, and the gramophone. Years of theatrical entertainments had also primed the public's appetite for the new world of moving pictures. As the Industrial Age economy expanded and cities grew during the late 19th century, entrepreneurs fed the emerging middle class' appetite for melodramas, vaudeville shows, and *tableaux vivants* (stage depictions of historical events with motionless actors). Railroads and canals enabled theatrical producers such as Dion Boucicault to bring elaborate stage shows to the American heartland.

Storytelling

At first, moving-picture audiences were eager to watch whatever had been photographed. Brief clips of slapstick clowns and vaudeville dancers provoked laughter; footage of speeding fire engines and trains caused shrieks upon first viewing. It wasn't long, however, before audiences bored of the simple novelty of seeing pictures move. Vaudeville theaters, a major venue for moving picture shows, relegated them to last place on the program.

Fortunately, other entrepreneurs still sensed opportunity in this infant art form. One of these was the brilliant French magician Georges Méliès, who in 1895 owned the Théatre Jean Robert-Houdin and realized that Lumière cameras and projectors could be used to make better illusions than magic lanterns ever could. Film historians cite Méliès as the first person to discover how fundamental trick shots such as in-camera edits, double exposures, and slow motion could be used to create amazing effects. Most important of all, his films told stories. His 1900 short of the classic *Cinderella* tale was well received by audiences and he followed it with many more trick films, including *A Trip to the Moon* (1902), one of the cinema's earliest "hits."

In the United States, Edison employee Edwin S. Porter's 1903 films *The Great Train Robbery*, *The Life of an American Fireman*, and *Uncle Tom's Cabin* were major successes, and films began to be exhibited for five cents at music halls (or "nickelodeons"). Nickelodeons opened in many cities, their operators including such people as Thomas L. Tally, owner of the Electric Theater, in Los Angeles, and four Newcastle, PA brothers named Warner who would later establish what is still one of Hollywood's greatest studios.

As the Nickelodeon Age blossomed and filmmaking and exhibition proved itself a profitable business, innovators such as D.W. Griffith, Charlie Chaplin, Carl Laemmle, Mack Sennett, and Cecil B. De Mille came upon the scene and creatively exploited the potential of cinematic storytelling. Griffith's 12-reel *The Birth of a Nation* was not only revolutionary for featuring the close-ups and long shots we all take for granted today, but its controversial content caused riots while also earning millions of dollars. It was the first movie ever shown at the White House and it prompted

President Woodrow Wilson to comment that it was "Like writing history with lightning."

By the 1920s, movies were a huge industry and a national obsession. Millions of people thrilled to the exploits of the Keystone Cops, *The Perils of Pauline*, and Rudolph Valentino every week at newly built movie palaces. At the same time that the cinema was moving out of seedy arcades and nickelodeons and into plush new theaters, it also was heading west to Hollywood for dependable sunlight and year-round shooting weather. American movies soon dominated the world market. Film camera and projector technologies steadily improved and by the end of the decade, synchronous sound had become commercially practical.

Technology Transition

Motion-picture sound had been experimented with since Edison, and although the Warner Brothers' 1927 film *The Jazz Singer* wasn't history's first "talking picture," its success with audiences marked a turning point: Sound had become an essential part of the movies. Warner's disc-based Vitaphone system was overtaken by electrical engineer Lee De Forest's competing inventions: the printing of an optical variable-density stripe directly on the film that "married" sound and picture and maintained synchronization, and the Audion vacuum tube to amplify sound in theaters.

Innovations such as color, wide screen, and stereo would follow in the decades to come, but the basics of motion-picture production and exhibition technology would remain fundamentally unchanged throughout the 20th century. Simply stated, film is a long ribbon of clear plastic, coated with photographic emulsion that is mechanically transported through a camera, where a shuttering mechanism exposes it to light, recording on the film a series of separate, sequential still images. Afterwards, the film is mechanically run through a processing bath, where the images recorded on it are developed. Eventually, the film is edited with a blade and finished prints are photochemically duplicated at a laboratory and distributed via trucking services to theaters. In the theater, the film is mechanically fed through another machine (a projector) for display on a screen.

Amazingly, the basic photochemical-mechanical technology of film is essentially unchanged since its invention more than a century ago. Film is a reliable, mature technology capable of providing a compelling entertainment experience. The motion-picture industry has long been a multibillion-dollar business as well as a social phenomenon and worldwide manifestation of humanity's continuing preoccupation with image-making and visual communication. Movies have been used not only to entertain, but also to persuade—sometimes with disastrous results. From the Marxist montages of Eisenstein to the Nuremberg nazis of Leni Riefenstahl to the varied political overtones of modern movies, powerful people know that the cinema is a proven means of winning hearts and minds.

The moving image is the most powerful form of communication ever devised. And the large-scale moving images of theatrical motion pictures have perhaps the greatest impact of all due to their sheer size and the intimacy that the cinematic experience offers audiences. Filling nearly the entire visual field and accompanied by multiple audio tracks, movies provide an engrossing entertainment experience.

Today, however, this most powerful of all image-making technology is entering a brand new phase in its existence, one that is more profound and potentially far-reaching than any other that has gone before. This new phase is the Age of Digital Cinema and the reason you're reading this book. But to begin to comprehend this new age, we need to again review some history.

Electronic Imaging

While Edison, the Lumières (Figure 1-5), Méliès, and others were planting the seeds of the movie industry, the evolutionary tree of the moving image sprouted a separate branch that would become television. And whereas film was based on mechanics and photochemical media, television has been—for the most part—purely electronic.

The terms *television* and *video* today are not necessarily synonymous. The word *television* can also infer broadcasting and receiving radio signals that carry video content. *Video*, however, tends to mean electronic image content. And video can be stored on various media—including magnetic tape, computer disk arrays, and optical discs—for playback on a variety of displays. Early in the 20th century, however, *television* and *video* tended to be synonymous terms.

Determining who "invented" television is as difficult a task as doing the same for film. Both motion-imaging technologies derive from common ancestry and both are the result of incremental advances made by many

Figure 1-5

The Lumière Brothers'
exhibition of motion-
picture films set the
stage for the coming of
the movie industry.
(© *Corbis.*)

people over a long period of time. Technology consultant and historian Mark Schubin has written about this subject over the years in the professional trade magazine *Videography*. In its April 1999 issue, Schubin cites the German inventor Paul Nipkow's 1884 patent application as what is today considered "the basic television patent." Schubin goes on to explain that Nipkow's patent "...described a mechanical device with a spinning, perforated disk that could reproduce scanned, electronic images." Schubin adds, however, that "well before Nipkow's patent application, technologists around the world had proposed television camera systems. A book was even published on the subject in 1880.

"Some schemes suggested a mosaic of tiles of light-sensitive material (usually selenium), each connected to an equivalent point in a light-emitting array," Schubin continues. "One such system, written up in the periodical *English Mechanic* in 1879, used hot, glowing wires for the display."

That invention was the creation of Denis Redmond, of Dublin, in 1878. Schubin states that Portuguese professor Adriano de Paiva wrote a proposal for a video system that same year. Television historian George Shiers' article "Historical Notes on Television Before 1900" (*SMPTE Journal*, March 1977), lists television "inventors" in chronological order as Redmond, Ayrton, Perry, Carey, Sawyer, Leblanc, Senlecq, Lucas, Atkinson, Nipkow, Bakhmet'ev, Weiller, Sutton, Brillouin, Le Pontois,

Blondin, Morse, Majorana, Jenkins, [Nystrom], Szczepanik, Dussaud, Vol'fke, and Polumordvinov.

And Schubin adds that "...it took more than 40 years after Nipkow's patent before a video camera could deliver a recognizable human face, but that camera, like Nipkow's proposal, used a perforated spinning disk as the scanning device. Cameras with spinning disks or mirrored drums came to be known as *mechanical* video cameras after the alternative, the electronic video camera, was introduced."

That first successful camera was the creation of Scottish inventor John Logie Baird, in 1925. Although mechanical and all-electronic television coexisted for a time, all of the world's television systems today are electronic. Other notable contributors to the evolution of television include: Boris Rosing, inventor of the first picture tube, in 1907; Philo T. Farnsworth, possibly the first person to send an image from an electronic video camera to a picture tube, in 1927; and Vladimir Kosma Zworykin who refined the sensitivity of the picture tube, making it practical for television, in 1930.

A New Beginning

The advancement of television was interrupted by other technology imperatives such as radar and sonar during World War II. In the postwar era, the small screen became a standard household appliance. Like the movies, television is principally a storytelling medium, but its commercial advertising also made it a major force in world commerce; its news broadcasts were vital in time of war, and these shape public opinion to this day. Television's impact has been the topic of scholarly analysis for decades, beginning most notably with Canadian educator Marshall McLuhan in the 1960s.

During the 20th century, film and television technology evolved separately for the most part, although their histories did intersect. Film became a major origination medium for television from the 1950s onward. Theatrical movies also have been an enormous source of content for television. The reverse was not, however, the case.

Early attempts at electronic cinematography—the use of video as a substitute for film in theatrical movie production and exhibition—were unsuccessful. There simply wasn't enough picture information ("resolution") in television images to enlarge it for the big screen. Also, the structure of the television image—which runs at a different frame rate than film and involves the "interlacing" of each "frame" of video (see Chapter 2)—made the use of video as a medium of theatrical film origination impractical for many years. Even with the invention of the first commercially successful videotape recorder—by the Ampex Corporation, in 1956—it would still be at least 20 years until video could be edited as easily as film with its clearly visible, separate picture frames.

Harlow, a 1965 biography of the troubled starlet, suffered not only from the limits of the "Electronovision" video-to-film conversion technology available at the time, but from a competing movie with the same title released almost simultaneously. Whereas television programs photographed on 35mm occasionally found their way successfully to theater screens, the resolution of video-originated images (whether captured on film as a "kinescope" or on videotape) lacked a sufficient level of picture quality to satisfy audiences of large-screen movies.

The motion-picture industry had also experimented with using television technology for exhibition at least since the late 1930s. Schubin writes (in *Videography*, June 1990) that "…the czars of large-screen motion picture entertainment in Hollywood had taken note of the advent of television. The Academy of Motion Picture Arts and Sciences (AMPAS, the Oscar-awarding organization) created a committee to study television in 1938 and came to the conclusion that motion picture theaters should install giant television screens. Paramount Pictures hedged its bets and invested heavily in both home television (it had a financial interest in four of the first nine U.S. TV stations and in DuMont) and in theatrical television, for which it created an 'intermediate film' system.

"The intermediate film system, crazy as it may seem today, actually worked. Signals were transmitted to theaters, where they were recorded on film; the film then passed through a rapid processing system and was projected 66 seconds later."

Schubin's article goes on to describe Scophony, a British company with a similar electronic theatrical exhibition system. These and other early schemes ultimately proved to be more trouble than they were worth, impractical alternatives to the much simpler business model of shipping film to theaters and projecting it with excellent results.

Large-screen electronic projection gradually improved over the decades, but by the 1960s, it was still cost-prohibitive and not really practical for theatrical use. Relegated to specialized applications such as the military, which used giant *Dr. Strangelove*-style graphic displays to view air-defense radar data, electronic projection employed high-brightness television-style cathode-ray tubes (CRTs) or the oil-based Eidophor "light valve." It was in this defense realm, however, with its generous R&D funding, that solutions would emerge to revolutionize electronic theatrical projection.

As Edison and his contemporaries knew, photography and projection were the cinema's two major components. A century later, both would be transformed by a third technology born not of film or television, but rooted in managing information. And when this technology was applied to managing *picture* information, it would lead to the digital cinema revolution. That technology is the computer.

Computer Graphics, Digital Video, and High-Definition Television

Like image making, humans have been counting and calculating for millennia. Calculations required for agriculture and trade gave birth to arithmetic and early inventions. There were calendars to reckon time, the abacus for counting, and the astrolabe to chart celestial movements for navigation. The use of punch cards by 18th century Frenchman Joseph-Marie Jacquard's automatic pattern-weaving textile loom helped inspire Englishman Charles Babbage's idea for the Analytical Engine, an early mechanical computer. In time, the path of computer evolution diverged from that of mere calculation; a *computer* could store an instruction "program" and hold and retrieve data automatically.

World War II catapulted computer research, with machines such as Colossus and ENIAC developed for code-breaking and ballistics research. By the 1960s, *transistor* and *IBM* were household words and computers had become essential to government and industry. As an information technology, computers eventually became image-making devices as well, harnessing the information power inherent in pictures. Early applications included using computers to drive the movements of mechanical engineering "plotters," which drew images onto paper for engineering studies. In 1962, a graduate student at the Massachusetts Institute of Technology named Ivan Sutherland authored a doctoral thesis (titled *Sketchpad: A Man-Machine Graphics Communication System*) that enabled anyone to use a light pen and a row of buttons to create basic images. In 1964, Sutherland and Dr. David Evans, of the University of Utah, developed the first academic computer graphics (CG) department. There and elsewhere, a long series of innovations continually improved CG technology, including the display processor unit, for converting computer commands and relaying them to a picture tube; the frame buffer, a digital memory to temporarily store images; and the storage-refresh raster display that made CG picture tubes practical.

The use of computers to calculate and render complex information visually proved beneficial to solving engineering and scientific tasks and brought about the field of computer-aided design (CAD), a vital tool of business and industry. CG imaging advanced during the 1960s as a part of computer science, evolving in the labs of universities, governments, and, especially, the aerospace industry. The U.S. space program was not only the impetus for the miniaturization of computer electronics that could fit onboard space capsules, it also provided the impetus for computer technology to be applied to video. This would prove an important contributing factor to the eventual arrival of the digital cinema revolution.

Digital and Analog

The first live video transmitted from the moon presented NASA with a challenge (Figure 2-1). Timing is crucial to video, which will not display a stable picture if its many complex signals are not properly timed. The 250,000 miles that the live, microwaved video traveled on its way back to the earth during the Apollo lunar exploration program introduced a synchronization problem. The solution in 1969 was to take advantage of a type of computer technology known as *analog-to-digital* conversion. This was used to turn the video—which was "analog"—into "digital" data, synchronize those data in a computer frame buffer, and then output them as stable time-base-corrected analog video acceptable to the world's TV transmitters and broadcast systems. The use of the digital computer technology was like correcting the off-speed sound of a warped phonograph record with the regularity of a metronome.

And this brings us to the meaning of the word *digital*, part of the title of this book. It's a word seen frequently in today's world, a product of the computer technology so essential to communication, impacting everything from home video DVD discs, to cell phones, to personal computers, and the Internet. Most important for our purposes, *digital* also defines the cinema technology that's revolutionizing the way movies are made and exhibited.

Simply stated, *digital* is the way in which modern computers process information. Basic in concept while complex in execution, digital process-

Figure 2-1

Live video of Apollo 11 astronaut Neil Armstrong descending the Lunar Module *Eagle*'s ladder on July 20, 1969. The digital technologies that helped make these motion images possible not only documented the human race's greatest technological achievement, they also eventually led to the digital cinema revolution.
(*Image courtesy of NASA.*)

ing reduces all information—including pictures—to a series of 0's and 1's. This is done because digitized information is a convenient way for computers to process, copy, and edit. And when it comes to video and CG, it typically results in higher-quality images than are possible with *analog* technology. Think of a clock with hands as being an analog timepiece, and a clock that displays only numbers as digital.

Film and nondigital video (such as VHS tape) are described as *analog* because they are physical representations of the real images they represent. Film is analog because its series of separate photographic transparencies or *frames* are an analog of the light reflecting off the real-world things (people, trees, etc.) the film was exposed to. Analog video is based on what are called *continuously variable signals*, fluctuating voltages stored on magnetic tape or some other media. These changing electrical voltages mirror a continuously varying wave. (The light that your eye or a camera sees is also a continuously varying wave.) In video, increases or decreases in the height of the wave (voltage) and distances between wave peaks (frequency) convey picture or sound information.

Digital (which simply means *numerical*) recording takes periodic *samples* of information along that wave and assigns (*quantizes*) the information into a series of 0's and 1's. This *coded* information is then divided up into *bits*, one eighth of which make a unit known as a *byte*. A *megabyte* is one million bytes. Needless to say, recording moving images requires the high-speed processing of an enormous amount of 0's and 1's, but that's what computers (and the microprocessors that make them run) are designed to do. Sometimes portions of this huge amount of information can actually be discarded in a process known as *compression* (see Chapter 4) with no harm to the digital picture or sound information.

Digital is better than analog because digital signals are not subject to the distortion or degradation of analog signals when they are processed or reproduced. Each time an analog signal is re-recorded (as in editing) there's an increase in signal noise that degrades the picture. If you have ever copied a VHS tape, you have seen how the picture quality degrades. In film as well, when special-effects shots require double exposures, visual film "grain" noise is increased, which is why many of the trick shots in older science-fiction films look obvious.

In digital signals, however, a 1 is a 1 and a 0 is a 0. There is no "gray area" in between to introduce degradation. The 10th or 20th or even 30th time a digital image is copied looks as good as the first. This is why copies of digital tapes are sometimes referred to as "clones." Image quality is greatly improved, just as a digital audio CD lacks the *clicks* and *pops* of an analog LP phonograph record. Digital imaging—whether as CG or digital high-definition (HD) video—provides filmmakers and film exhibitors with many powerful new creative options and economic opportunities, as we shall see.

The preceding explanation of digital and analog is extremely basic. For a more thorough understanding of what *digital* means, a wealth of information exists, including entire books written on the subject. A good introductory tutorial, by technology consultant Mark Schubin, appears in Chapter One of *Creating Digital Content* (McGraw-Hill, 2002).

For the filmmaker's purposes, the inner workings of digital technology should ideally be all but transparent (Figure 2-2). More important is understanding the cinematic craft and how to use production tools to achieve it. One need not know the inner workings of internal-combustion engines in order to drive a car. What's important for filmmakers are the cinematic arts, or as digital movie innovator and director George Lucas explains:

"The medium that I'm in and that most people are working on at this point is the storytelling medium. For those of us who are trying to tell stories, the agenda is exactly the same. It doesn't make any difference what technology you shoot it on."

Figure 2-2

Digital synchronization made live video of NASA's lunar missions possible. The long-term impact of this innovation would improve electronic imaging immeasurably. (*Image courtesy of NASA.*)

Video and Film

By the late 1960s, computer-generated animations of pulsating patterns by such artists as Mark, Michael, and John Whitney Jr., and Kenneth Knowlton pointed the way toward a new kind of "expanded cinema." Korea's Nam June Paik altered the time base of video signals to create "video art." Film director Stanley Kubrick and budding special-effects wizard Douglas Trumbull employed computer technology for the slit-scan photography of the "star gate" sequence appearing at the end of the 1968 film *2001: A Space Odyssey*.

Companies and engineers that had provided digital solutions to NASA's lunar-video needs turned their research into innovative new products for the television industry during the 1970s. Digital frame synchronizers made live TV station location reports more reliable. Digital time-base correctors made the new cassette-based video formats broadcast compliant. Digital character generators made onscreen weather and sports scores much more readable and attractive. And digital effects systems made "page-turns" and other image transitions in television possible. But television wasn't the only moving-image area advancing because of computer technology.

"Our goal was to be able to use video as an acquisition and mastering tool to make 35mm motion pictures," recalls Ken Holland, founder and

Figure 2-3

Ken Holland, Founder and President of IVC (International Video Conversions) High Definition Data Center (Burbank CA) is a pioneer in applying electronic technologies to theatrical motion picture production. (*Photo by Dick Millais, courtesy of IVC International Video Conversions.*)

president of IVC (International Video Conversions) High Definition Data Center, in Burbank, California (Figure 2-3). Holland came from a defense electronics and TV-engineering background, and in 1970 cofounded a company named Image Transform; its image-enhancement and visual-noise reduction processes were among those used to improve live video from Apollo lunar missions.

"It was an uphill battle because of television's frame rate and limited resolution," Holland continues, describing his work with the IMAGE 655 system, which was used to make theatrical feature films. "We had to do a lot of image processing to closely match a 35mm quality look. This was actually an extension of 525-line video, with 130 more lines per frame and the frame rate slowed from 30 to 24 fps. It was the first practical electronic motion-picture system suitable for a 35mm theatrical release, and it provided adequate quality and production advantages for nonscope films shot primarily on a stage or interior locations. We were fairly successful with IMAGE 655 during the mid 1970s to early 1980s. It was used to tape original production footage and/or create special effects that we converted to 35mm for a large number of Hollywood features of the period."

Those feature films included *Norman, Is That You?* (1976), *Monty Python Live at the Hollywood Bowl* (1982), and effects shots for the 1978 version of *Heaven Can Wait*. Other Image Transform innovations Holland supervised included the Image Lab, a custom-built motion picture laboratory, and Imagematte, a system for superimposing actors on fake backgrounds similar to what TV weathercasters use, but of sufficient image quality to be transferred to film.

Holland's references to *resolution*, *525-line video*, and *30 to 24 fps* provide an opportunity to further define key terms in both video and digital cinema. *Resolution* (also known as *definition*) refers to the amount of detail in a picture. Generally speaking, film has more resolution than video. Television and video in North America is based on the National Television System Committee (NTSC) standard of 525 lines (approximately 480 of which are used to comprise a picture). The Phase Alternation by Line (PAL) television standard (used in the United Kingdom, Germany, and many other countries) is based on 625 lines of picture information (575 or 576 of which comprise a picture).

Most video employs what is known as *interlace scanning*, which, in NTSC video, divides each of the 30 *frames* (separate still pictures) of video that are displayed every second into two *fields*. Each field contains half the number of lines necessary to make a picture. First all the even-numbered lines are displayed, then all the odd-numbered lines. This is done to conserve space in broadcasting video signals. When all 60 fields (30 frames, each one seen twice) are displayed every second, the eye perceives smooth motion.

Computer screens, however, display their video *progressively*, that is, each frame is scanned onto the screen and shown in its entirety before another frame is displayed. Progressive scanning makes better quality pictures than interlace scanning does. Film also displays one entire frame at a time. Filmmakers prefer progressive scanning because it eliminates the picture degradation caused by interlacing.

Another difference between motion-picture film and video is that film displays 24 separate frames per second (fps), a speed standardized in the late 1920s when sound was introduced to the movies. The speed of 24 fps was chosen because it rendered motion well, and was the slowest speed at which optical soundtracks could still reproduce acceptable audio. When films are shown on television, an interpolation process known as *3:2 pull-down* is employed to adapt the 24 fps speed of film to the 30 frame-per-second/60 field-per-second refresh rate of video.

The Pixel's Premiere

Image Transform wasn't the only organization during the 1970s doing research on using higher-resolution video for shooting theatrical movies. Nippon Hoso Kyokai (NHK, the Japan Broadcasting Corporation) was working on what it called HDTV, or high-definition television. This video system featured 1052 lines of resolution and also included cinema-style widescreen images with a width-to-height ratio of 16 by 9, as opposed to the 4 by 3 *aspect ratio* of regular NTSC TVs. HDTV was not only a new recording video format, it was also an entire proposed broadcasting standard. It would be in development for many years and was not originally intended for use in making theatrical movies.

CG also continued to advance during the decade, with 1974 witnessing the birth of the Association of Computing Machinery's (ACM's) Special Interest Group on Computer Graphics (SIGGRAPH). The organization's various local chapters and national conferences provided a forum in which CG experts could share information. Scientific imaging (animations of weather patterns, architectural "walk-throughs," wind-tunnel airflow simulations) was characteristic of early CG research. Eventually, CG practitioners found they could effectively showcase their latest advances in the form of short, cartoon-style animations. These clips conformed to the familiar paradigm of cartoons while also demonstrating how CG could replicate the visual dynamics of the physical world.

Drawing upon mathematics, physics, and a multitude of other sciences that measure and quantify the way that matter appears and behaves in the physical world, CG scientists, software writers, and designers/animators continually worked toward advancing computers so they could generate "photorealistic" imagery that displayed three-dimensional perspective. This long process of discovery went from inventing ways of drawing wire-frame "skeletons" of objects, to developing algorithms for what's known as *polygon rendering* to give them surfaces (metal, stone, skin, fur, etc.), to animating them in ways the human eye finds appealing. Improving and refining this process continues to this day.

The basic unit of CG is the picture element, or *pixel*. Digital, random-access computer memory (RAM) determines the number of pixels on the screen, their location, color, etc. Each pixel is assigned a particular number of memory bits in a process known as *bit mapping*. The more bits, the greater each pixel's potential color combinations. *Bit depth* is a term that refers to the number of these bits. The greater the bit depth, the more color levels possible. In a 24-bit system, the computer provides for eight bits each of red, green, and blue. It is from these basic colors that all other hues are created in CG. A 30-bit system provides for 10 bits per basic color; a 36-bit system, 12. More color means more shading, detail, and—in the hands of a skilled artist—realism.

As processing power and memory capacity advanced, computers gradually shrunk in size, and the technology proliferated in a wide variety of industries. George Lucas' 1977 *Star Wars* proved a landmark in the use of computers in filmmaking, but only insofar as they were used to automate motion-control and lens-adjustment systems used to repeat the movements of spaceship models in order to photograph convincing (albeit impossible) scenes of interplanetary battles and flight. Although computers didn't specifically generate the images used in the movie, they were integral to controlling its special-effects cameras. *Star Wars* won six Oscars, including the prize for visual effects, and made it clear to everyone the role that computers could now play in making movies.

By the early 1980s, CG was no longer the sole domain of university computer science departments. Lucas formed a Computer Development Division that would eventually become Pixar, which is famous today for such movies as *Toy Story*, *Finding Nemo*, and *The Incredibles*. Other companies such as Robert Abel & Associates, MAGI, and Pacific Data Images opened for business as driving forces in making CG practical for TV and motion pictures. In the meantime, computer technology continued to decline in price, and the first personal computers became available. Years earlier, Gordon E. Moore (physicist, cofounder, and chairman of Intel Corp.) had predicted continuing increases in computer power accompanied by continuing price decreases. Dubbed "Moore's Law," this phenomenon continues to the present day.

The 1982 Disney film *Tron* marked another milestone in computerized filmmaking; it contained more than 15 minutes of computer-generated images and live action composited with CG. Four CG companies collaborated on the film, which although not a critical success, was a dramatic demonstration of how a totally new look was possible using digital imaging tools. To transfer these computer images from picture tube to movie film, *Tron*'s producers called on the Constantine Engineering Laboratories Company (CELCO), which had been making special cameras for NASA and the defense community. These devices were used to print satellite image data into large photographic prints. CELCO continues to be a major digital film-recording technology developer and provider for the motion-picture industry.

More films followed that featured CG sequences, including *The Last Starfighter* (1984) and *Young Sherlock Holmes* (1985). Companies offering CG systems to television also opened for business, providing graphics systems that could be used to animate video of "three-dimensional flying logos" of TV station call letters. The Quantel Paintbox system, meanwhile, brought about the age of over-the-shoulder newscaster graphics that individual TV station art directors could customize. CG helped TV look more high-tech, and commercial producers began using it as well to create attention-getting imagery for products ranging from scrubbing-bubble bathtub cleaner to indigestion medicine. But CG's impact on the cinema was of much greater importance. An era had arrived in which any image the mind could conceive could now be rendered visually onscreen. And with sufficient photorealism, CG could be composited with live-action 35mm photography to create convincing imagery that might not be possible using previous "analog" methods.

Production, Post, and Projection

While CG was moving out of the academic world and into television commercials and movie theaters, NHK's HDTV technology was also advancing. *Godfather* director Francis Ford Coppola became the first Hollywood filmmaker to test out 1,125-line HD equipment in 1982. Although he didn't make a movie with it, he did outfit an "Electronic Cinema" van to experiment with video as a means of previewing film photography for his movie *One from the Heart*. At the time, Coppola commented:

"Video is a genie. Video can assume many forms and be handled like quicksilver. And that to me is almost more important for now. I know that with some research bucks thrown into it, video in the very near future will be able to equal the luster and quality of film."

Research bucks to achieve that luster and quality were indeed being spent, principally by NHK. In the meantime, other important components of what would become the digital cinema revolution were also falling into place. *Star Wars* director George Lucas' Lucasfilm introduced the EditDroid, which was one of the first motion picture and television "nonlinear" video editing systems. Instead of employing reels of film or recording tape, such systems rely on digital disks, which don't require rewinding or fast-forwarding; hence they are "nonlinear." Although an unsuccessful product, EditDroid would help pave the way for Avid Technology's Media

Composer several years later. Motion-picture film editors were slow to warm to the Avid at first, but by the mid 1990s, digital nonlinear editing had become the standard in Hollywood.

Elsewhere, high-quality digital projection was also becoming practical. During the 1970s, the Hughes Aircraft company's Malibu, CA research lab built large-screen displays for the command and control centers of U.S. Navy ships. The technology involved was known as the Liquid-Crystal Light Valve, which eventually was adapted to display computer graphics and full-motion video. In 1992, Hughes joined with JVC to commercialize the technology, renaming it D-ILA (for Digital Image Light Amplifier).

An historic moment occurred on June 18, 1999 when two D-ILA projectors were used to present George Lucas' *Star Wars Episode I: The Phantom Menace* in theaters on both coasts. And on that same date, two additional theaters showed the movie digitally using a competing technology, DLP Cinema, from Texas Instruments (see Chapter 16).

Although projected digitally in these historic demonstration screenings, *Star Wars Episode I* had been photographed on 35mm film, as all movies have been for nearly a century. This, however, would be Lucas' last movie to be shot on film. The maverick moviemaker had already decided that digital imaging technology and HD video had progressed to the point where a major feature film could be photographed entirely in the digital domain, and he would be the one to do it. Lucas and his Lucasfilm studio set their sights on making digital filmmaking history and the Sony Corporation—for many years an innovator in professional and broadcast-grade video camera and recording technology—saw their initiative as a market opportunity to become the "Kodak" of the 21st century. The collaboration between Lucas and Sony would make history and signal the start of the Age of Digital Cinema.

Lucas Takes the Lead

It's hard to say exactly when the digital cinema revolution began in earnest, but George Lucas' 1996 letter to Sony's research center in Atsugi, Japan probably provides the best historical starting point.

Lucas had long been a supporter of developing new technologies to enhance filmmaking creativity and workflow. A technology-savvy writer/director/editor/producer, his use of computerized motion-control and new sound-effects techniques on the original *Star Wars* (1977) set new standards for cinematic innovation. The success of that Oscar-winning film and its sequels provided him with the resources to invest further in new filmmaking technologies. In 1982, his Lucasfilm company developed the EditDroid—one of the first nonlinear video editing systems. Lucas was also an original founder of Pixar, which pioneered feature-length movies made with computer-graphic imaging (CGI). His Industrial Light + Magic (ILM) subsidiary of Lucasfilm Ltd. has won 14 Oscars for its digital special-effects work on some of the most successful movies of all time. Lucasfilm has also applied its digital expertise to television projects such as *The Young Indiana Jones Chronicles* and to an educational software division. Lucasfilm's THX initiative, meanwhile, has a long history of working with theaters and home-video distributors to improve sound reproduction.

High-definition video had already been used to make theatrical movies by the time Lucas wrote his letter. Such films included director Peter Del Monte's 1987 *Julia and Julia*, which cinematographer Giuseppe Rotunno photographed using Sony's analog High Definition Video System (HDVS) cameras and videotape recorders. Because HDVS was an analog system intended for broadcast use, however, its 30-frame/60-field interlaced recording format produced substandard images when converted to motion-picture film for theatrical exhibition. Unlike video for television, motion-picture film frames are *progressive*; they are displayed one at a time, as opposed to being split into even- and odd-numbered lines and displayed in "interlaced" fashion. Also, movie film is displayed at a rate of 24 frames each second. The interpolation required to convert 30-frame/60-field interlaced video to 24-frame-per-second 35mm film typically produces image "artifacts" that audiences find distracting.

Knowing of Sony's years of HD R&D and its recent development of digital HDTV cameras and recording (Figure 3-1), Lucas' 1996 letter asked if the company could make an HD system that records, stores, and plays back images at the film standard of 24 frames per second. In this way, each frame of HD video could be transferred to film without the "3:2" interpolation required for transferring 30-frame/60-field video to 24-frame film. In addition, Lucas requested that the new Sony digital HD system use progressive scanning (which computer graphics displays also employ) so that interlacing would cease to be a problem.

Figure 3-1

Digital cinema pioneer George Lucas shooting *Star Wars Episode II: Attack of the Clones* with Sony/Panavision HDW-F900 camera on location in Italy. (*Photo by Lisa Tomasetti, © Lucasfilm Ltd. & TM. All Rights Reserved.*)

Lucas' request made perfect sense, given his needs. As a filmmaker who already integrated a large amount of "synthetic" computer-generated imagery into his movies, eliminating film entirely from his production process would ultimately save money and speed up production workflow. Shooting live-action scenes in 35mm required scanning each film frame into digital data for integration with *Star Wars*-style digital imagery. But if he shot his live-action scenes in a digital HD format free of the technical impediments of broadcast video formats, integration with digitally generated computer imagery would be streamlined.

Lucas had already tested Sony's standard-definition Digital Betacam format for selected shots in his 1999 *Star Wars Episode I: The Phantom Menace*. He liked what he saw. "We had some test shots that we captured and inserted digitally and nobody could tell which were which," he revealed in a 2001 interview published in *Creating Digital Content* (McGraw-Hill, 2002).

Sony rose to Lucas' challenge and brought forth its CineAlta format, which recorded 24-frame progressive-scanned digital HD video. Next came the need for quality cinematographic lenses.

"Panavision was very important," Lucas said, referring to a leading motion-picture camera, lens, and production-equipment rentals company. "The first road block we came to was, 'What about the lens?' So we went around to all the lens manufacturers, had a very hard time getting anybody to do it. [We] finally connected with Panavision, which was very enthusiastic…[we] got Panavision and Sony together, and they were both willing to make huge financial commitments in this area. And, you know, we consulted with both of them and sort of became the matchmaker."

In 2000, Panavision unveiled its advanced Primo Digital lenses for digital cinematography, mounting it on Sony's HDW-F900 digital 24p CineAlta cameras and further "Panavising" the camera with additional cinema-style modifications (follow-focus, extended viewfinder, matte boxes, etc.) familiar to filmmakers. Panavision then formed a new company with Sony Electronics to support this integration. By the time Lucas completed his all-digital *Star Wars Episode III: Attack of the Clones*, other directors had already made all-digital films with the CineAlta equipment. But Lucas' goal wasn't to be the first. He summed up his satisfaction with the CineAlta system before he'd even finished editing the movie in an October 25, 2000 ad in the trade paper *The Hollywood Reporter*. "We shot *Star Wars Episode II* in 61 days in five countries in the rain and desert heat averaging 36 setups per day without a single camera problem. We have found the picture quality of the 24p Digital HD system to be indistinguishable from film.

"Thank you Sony and Panavision for helping us meet our goal of making an entirely digital motion picture."

Six months later, during Sony's press conference at the annual National Association of Broadcasters convention, in Las Vegas, Lucas elaborated on his experience with digital filmmaking, and made the historic statement, "I think that I can safely say that I will probably never shoot another film on film. It's the same issue with digital editing. I've been editing digitally for over 15 years now, and I can't imagine working on a Moviola again."

Lucas also commented that he saved $2 million using digital tape instead of film, and that he was able to work much faster.

George Lucas' Digital Vision

During March and April of 2001, while still deep into postproduction on *Star Wars Episode II*, George Lucas took time out for an interview with the author of this book and with John Rice, coeditor of *Creating Digital Content*. What follows is the combined and edited result of those conversations.

Can you give us a sense of where you think the entire entertainment community is right now, in terms of digital?

Obviously, the video industry has gone digital faster than anybody [thought it would]. The difference between tape and digital tape is very subtle. I think that is going to transition reasonably fast, especially on the production side. And in film, it's going to be a slow process. We went through this when we developed nonlinear editing and everybody was pooh-poohing it and fighting it, and they did it for 10 years. And then, after about 10 years, we sold it to Avid, and it was still another three or four years before people started using the system.

What about digital as a creative tool; will digital image acquisition or digital postproduction change the way people create movies?

Well, it definitely changes the way you create on a lot of different levels. I think it's as profound a change as going from silent to talkies and going from black and white to color. It gives the artists a whole range of possibilities that they never really had before.

In this particular case, the possibilities center around malleability. They [digital tools] are much more flexible, in terms of what you can do with the image, how you can change the image and work with the image. So, I think that's the biggest issue the filmmaker will be faced with. The equipment is easier to use and at the same time, it allows you to get more angles, and do more things than you'd normally be able to do. And then once you've captured the image, the digital technology allows you to do an unlimited amount of changes and work within a lot of different parameters that just were not possible with the photochemical process.

Will having used digital make your final product a different film than it would have been if it were shot on 35mm?

The medium that I'm in—and that most people are working on at this point—is the storytelling medium. For those of us who are trying to tell stories, the agenda is exactly the same. It doesn't make any difference what technology you shoot it on. It's all the same problem. It's just that telling stories using moving images has constraints. What this does is remove a large number of those constraints.

Are constraints creative, time, financial?

There are all kinds. Yes, the constraints are time and money, but most importantly, creative. Now we can do things that we just couldn't do before in the special-effects business. When we moved into digital about 20 years ago now, we kept pushing things forward and were able to do things like the morphing sequence in *Willow* [1988], the water creature in *The Abyss* [1989], and—finally—the dinosaurs in *Jurassic Park* [1993]. You just couldn't do those kinds of things in an analog medium. It just was not possible, no matter how you thought of doing it.

Is the 1080-line 24p HD you used good enough, or would you like to see more resolution?

Well, we always want to see more resolution. You've got to think of this as the movie business in 1901. Go back and look at the films made in 1901, and say, 'Gee, they had a long way to go in terms of resolution. They had a long way to go in terms of the technology.' That's where we are right now and it's great because just think of what's going to happen in the next 100 years. I mean it's going to go way beyond anything we can imagine now.

With film, quite frankly, as a medium, the physics of it can't get much further than it is right now. We don't live in the 19th century, we don't live in the sprockets and gears and celluloid and photochemical processes like we did then. We've advanced beyond that. We are in the digital age, and [as far as] all the arguments about 'It changes how you do it,' of course it changes how you do things. Digital makes it easier for you to edit. Film is...a hard medium to work in. You've got to push real hard to get it to go anywhere. But once it's digital, it's...light as a feather.

The Internet has been full of rumors that you shot digital and film side by side. Was there any film used to shoot* Star Wars Episode II*?

There was absolutely no film at all used in the shooting of Episode II. We did have probably a case of film and a VistaVision camera with us, but to be honest, the reason was more of an insurance issue than it was the fact that we ever planned to use it. The insurance company kind of didn't 'get its head around digital' as fast as we'd hoped, and it said, 'We will lower your rates if there's a film camera on a truck somewhere that you could get hold of if you had a problem.' But we never had to use it.

We never had a problem with the digital camera and we shot under unbelievably difficult conditions. We shot in the desert. We shot in the rain. Sony cameras have been used all over the world for newsgathering under...battle conditions. And they survive, they get the news. And this [CineAlta] is now just a hybrid of that kind of thing. There's no reason why it can't work in the jungle, the desert, the rain. We had no problems.

You can have rumors forever. But it doesn't make any difference because we did shoot the whole thing on digital. We had no problems whatsoever with the cameras or with the medium or anything. The film looks absolutely gorgeous. You can say whatever you want. Eventually, the film will be out there in the theaters and you can judge for yourself, but I guarantee that 99 percent of the people that see the film will simply not know that it was shot digitally. Then there's the one percent who are the technophiles or the people who are the die-hard film people, as opposed to cinema people, and they'll say it's not real film. And you know, it's *not* real film; what we're talking about is *cinema*, which is the art of the moving image. It's very similar to film, but it's not the same thing.

Why do you think so many negative rumors circulated about your use of digital cinematography?

I don't think anyone likes change. The same thing happened when we created EditDroid. When we created EditDroid, nobody wanted anything to do with it. There were all these rumors that directors would take forever because they had so many more choices. It breaks down all the time. There were all kinds of reasons about why it wouldn't work. Well, it *does* work. You can argue it to death, but it works and it's better and eventually people will use it because why would you work with a push mower when you can have a power mower? It just doesn't make sense.

Let's talk about digital delivery; there's been talk that you're hoping a good number of theaters are ready to do digital projection.

Well, we're hoping. When we released *Phantom Menace*, we were able to release it in four theaters digitally. Then our other San Francisco company, Pixar, pushed it to almost 20 theaters on *Toy Story 2*.

To what end; what's better about digital projection?

The quality is so much better. I've spent a lot of my time and energy trying to get the presentation [of movies] in the theater up to the quality that we have here at the [Skywalker] Ranch or what you see in private screening rooms. That's why THX was invented and that's why the TAP [Theater Assistance Program] was invented. We've worked very hard to try and improve the quality of presentation at the level of the audience actually seeing the movie. It's a rather thankless job, but it's something that does need to be done. And we've been pushing digital projection for years and years now, and working with all the vendors to try to improve it and try to get it working in the marketplace. And now the business models are coming together to try to actually take the next step, which is, 'How do you pay to get the thing into the theaters?' And THX is very active in that whole area.

Digital far and away improves that presentation. You don't get weave, you don't get scratchy prints, you don't get faded prints, you don't get tears—a lot of the things that really destroy the experience for people in the movie theater are eliminated.

Is digital cinema projection technology there?

I think the technology is definitely there and we projected it. I think it is very hard to tell a film that is projected digitally from a film that is projected on film. If you have brand-new prints, after a few weeks, it's very easy to tell which is which.

Do you foresee your content being repurposed repeatedly for other-than-large screens?

I think ultimately, when broadband comes, that there's going to be a lot of entertainment that will be run through the Internet, and I think that's going to be a big factor in the near future.

Do you have that in mind when you're creating films or are you still producing for the big screen?

I'm producing for the big screen. Whether broadband Internet actually happens is still up in the air. We haven't gotten that far yet. I think content that will be produced for the Internet will have to be of a different caliber, because the kind of a revenue stream that's going to be involved is still up in the air. There could be Napsters out there, and if so, it's going to be very hard to create for the Internet because you're going to have to do it for ten cents. You'll be basically giving your content away because there's no revenue stream coming in, and all you can do is hope to sell T-shirts or something to make money. So, you might be able to charge a little bit, but that whole issue of how you finance the product on the Internet is still completely unknown.

Earlier you mentioned that the transition to digital creation of motion-picture content is comparable to the transition of silent to sound films. Is there an overall digital distribution revolution going on? If we came back here in 10 years, would we be talking about an entirely different world of entertainment?

I think what's going to happen is that the invention of digital, in terms of distribution, is going to open up the marketplace to a lot more people. It's going to be much more like painting or writing books, and it will take cinema in that direction, which means it's accessible to anybody. That will change a lot about how we view the entertainment business. Right now it's very tightly controlled by a very small group of individuals who make all the decisions about what is good and what is bad and what people should see and all that sort of thing. This is going to pretty much knock that away so that filmmakers from anywhere and everywhere will have access to the distribution medium.

Entertainment options are expanding; you can watch feature films on a laptop or a PDA. What do you think that does to

the accessibility or the quality of the content that's going to be available to the consumer?

The technical quality will continue to improve no matter what medium is being used, and especially now, with digital. We're sort of at the very bottom of the medium. In terms of digital, we're working in very crude terms with what we're going to be able to do in the next 100 years. But in terms of the creative quality of the product, you have two forces at work. One is that you're going to infuse a lot of fresh blood and fresh ideas and exciting new people into the business. But on the other hand, you're going to have far fewer resources to create what you're going to create. So those may even themselves out. All in all, I think that visual entertainment, movies, and that sort of thing will pretty much be the way they always have been, which is about 10 percent is really good, 20 percent is sort of okay, and the rest of it is terrible.

Is there a technology or product missing right now that you wish were out there? Is there something you've dreamed of?

No. In terms of what I am really dreaming about right now, I want to get a totally integrated system where it's all digital and we're all working in one medium. The biggest problem we've been having in the last 15 years is the fact that we have bits and pieces from different kinds of mediums, and we're trying to *kludge* them all together into one product. I think things will be infinitely easier and cheaper and much more fun for the creative person when we're working in one medium.

Are there any words of encouragement or warning that you think need to be delivered to the filmmaking marketplace?

Well, the one thing that we have documented without any doubt in the film business is that people like quality. They like quality sound and they like quality images. They like to look at something that really has a technical polish to it. They respond to that. They did it when we released *Star Wars* in 70mm. We did it when we started THX theaters.

Could you elaborate on the aspect of digital allowing you to be more creative?

Digital is just a different storage medium, a different capture medium. It doesn't change what I do at all. You know, it's lighter cameras, easier to use, and a more malleable image. I can change it and do things that I want to do, but they're all technical things, so it's just a more facile medium than the photochemical medium. I think that is the all-encompassing, over-riding factor in all of this.

The example I always use came from my friend, [sound designer] Walter Murch. He described the period we're going through as being similar to the period in the Rennaissance when artists went from fresco painting to oil painting. When you had to do frescoes, you could do only a very small portion at a time; the plaster had to be mixed by a craftsman. The colors had to be mixed by other craftsmen, and the science of mixing those colors so that they would dry just the right color was all very important. And then everything had to match. Each section that you did had to match with the next section because you were doing what was, in essence, a large mosaic. You had to paint very fast because you had to paint before the plaster dried, and you pretty much had to work inside a cathedral or something, a dark space lit by candles.

And it was all science. It was technology at its height in those days because the plaster had to be mixed just right. Florence competed with Rome; it was, 'Ooh, we've got a red that you can't believe.' If you study the history of colors, going from red and blue to mauve was a huge advance in art—in technology—every time they came up with a new color.

Today in film it's really the same kinds of issues. With oil painting, suddenly you could go outside and work in sunlight, work with paint that didn't dry immediately, continue to work with it. Come back the next day, paint over it, and that was a huge advantage to the artist in terms of his ability to change his mind and work in a medium that was much more malleable and much more forgiving. One person could do it. You mixed your own colors and pretty much that's the way the color dried. You didn't need a bunch of experts back there doing it for you. You didn't have to have the 15, 20, or 30 guys mixing paint, mixing plaster, building scaffolds, doing things that you would normally have to do if you were an artist.

And oil let you interpret the way light played on things. You would never have ended up with the Impressionists if it hadn't been for oil, because they couldn't do that kind of art in fresco. And so did it change the medium? Yes. Did it change it for the better or worse? That's a personal opinion. But at the same time, it allowed the artist to free up his thinking and to say, 'I want to go to a different place than I've been able to go to before.' And, you know, I greatly admire the fresco painters and I still think it's an incredible art form, but people choose their medium, and they choose it for the kind of work they do.

So that is the best description of digital. It's really going from the photochemical, fresco medium to the oil painting, digital era.

Did digital technology have an effect on location shooting?

The process on location where you're actually shooting digital is just fantastic because there are a lot of important things that happen. These things may seem very small to people that are not involved in the film business, but to those who are, these are very big issues. You don't have the situation of running out of film every 10 minutes. This means that you don't get to a point in the performance when everybody is ready to do that really perfect take, and somebody yells, 'Roll out!' and you have to wait 10 minutes for a film reload while the whole mood falls apart. And then the actors have to go and build themselves back up again. That's very debilitating, that kind of thing. It seems small, but it's very big.

The issue of being able to move the cameras around very fast, the issue of being able to watch what's going on; you see your dailies on the set as you're shooting with the departments watching. So if the hair is messed up, if the wig shows, if the prop isn't there, all those kind of things that you'd normally see the next day in dailies, you're looking at them right now. You can make changes immediately. You don't have to go back and reshoot something. That makes a big difference. On *Episode II*, I had the same schedule and the same crew as *Episode I*, and I was able to go from 26 set-ups a day to 37 set-ups a day. I was able to shoot under schedule by three days, and a lot of this had to do with not having that extra hour at night watching dailies. Not having to do reloading and that sort of thing really does speed up the process in terms of the amount of work you can do in a day.

How about the operational aspects of the 24p digital CineAlta cameras? Everyone on the crew is used to certain ergonomics that film cameras have.

Most of the people who capture images are capturing them electronically. It is a very small group of people that actually still use film, because most of the images that are captured are captured for the broadcast medium, and most of the cameramen are using hand-held Betacams or some such camera. They are very used to the way it works. The only people who are having difficulty are the people who are used to big, heavy cameras with giant magazines on them. But once they get used to the nice, light, little cameras, they'll embrace them because it's not that difficult.

So eventually, we may not even be discussing how movies are made—digitally or on film?

I don't think so. Twenty years ago, we went through this exact same controversy with EditDroid and there was a huge backlash and a lot of fighting, a lot of rumors, a lot of craziness: 'It'll never happen, we'll never use it.' Well, now it's hard to find somebody who still cuts on a Moviola. Digital nonlinear editing is so accepted that the idea of not using a nonlinear editing system is unthinkable, and nobody talks about it now. Nobody asks, 'Did you cut this on a nonlinear editing machine, or did you cut this on a Moviola?' People don't even ask those questions anymore because it's obvious. Ninety-nine percent of the people are using nonlinear editing.

What do your peers think of you shooting movies digitally?

I have very close friends who say they will never, ever go digital until the last piece of film is used up and the last lab is closed. Then there are others—Jim Cameron, and a lot of the other directors—who have come through here to see what we're doing. Francis Coppola is also very enthusiastic about what we're doing and is amazed at what we've been able to accomplish in terms of images.

How does it feel to be a pioneer?

I don't think of myself as a pioneer. I've gotten asked these kinds of questions early on in my career because I've spent a great deal of money on research and development, and I've tried to push the medium forward. It's really because I want to make the process of making films easier for the artist, because I want to make it easier for myself to cut the films, to shoot the films. And because I am an independent filmmaker, I have to make sure that every resource I have is used in the best possible and efficient way. Sometimes advancing the technology a little here, a little there, does just that. It cuts your costs and it also makes it easier for the artists to realize their vision. I can think of a lot of crazy movies, but up to this point, I've always been hitting the concrete ceiling—or celluloid ceiling, I guess we should say—that has not allowed me to fulfill my imagination.

The main thing is I'm very happy now. I have always been pushing the envelope of the medium because primarily I want to get the best possible image and the best possible way of telling my stories. And I've always found myself bumping up against that celluloid ceiling, the technology that says, 'You can't go here, you can't go there, you can't do that.' Or the economic resources, such that, 'You can get there, but it's going to cost you an arm and a leg to do it.' And my feeling is that the artist really needs to be free to not have to think about how he's going to accomplish something, or if he can afford to accomplish something. He should just be able to let his imagination run wild without a lot of constraints. And that's really what digital is allowing us to do.

Crew Perspective

Shooting *Star Wars Episode II* was a learning experience not only for writer/director/editor/producer George Lucas, but also for his producer Rick McCallum, his postproduction supervisor and technical director Mike Blanchard, and for his principal engineer Fred Meyers. In the following interview (conducted by the author with Bob Zahn, Associate Member of the American Society of Cinematographers in March of 2001), the three relate their historic role in making the first complex, digitally photographed, effects-laden motion picture.

Why did you use 24-frame progressive-scan digital high-definition video to shoot Star Wars Episode II*?*

McCallum: Star Wars films are very complex and technologically challenging films to make. *Episode II* will have some 2,200 shots, and every single frame will have some kind of digital effect or animation comped into it. For us not to be totally in the digital arena is absurd. For us to be shooting film—sending it out to a lab, having it developed, having it sent back, then having it transferred to tape for editing—is ludicrous.

Can you quantify in terms of dollars what working in 24p HD means to your budget?

McCallum: Absolutely, but remember every single shot is a digital shot for us, which has a huge impact in terms of scanning in and scanning out of the computer. For us, the difference is about $1.8 million. We shot the equivalent of 1.2 million feet of film. That's the equivalent of 220 hours. When you take the cost of the negative, developing and printing it, the transfer, the sound transfers, and the telecine, it equals a serious amount of money. And if you're shooting in different countries, you have that negative shipped out, processed, and shipped back. There's freight agents involved and you risk your negative being lost or destroyed. When you're shooting in the digital arena, 220 hours of hi-def tape is $16,000.

And if you're simultaneously making a clone, which becomes your safety master, that's another $16,000. And then your down-conversion to put it into an editor is another $16,000. So for $48,000, you're making a movie without any of the cost of processing or transferring negative.

Did 24p HD save time?

Absolutely. The tape we shot in the morning was in our editing machine by lunchtime and our editor was cutting from the first day. That kind of speed is very important to us. Our goal was to make filmmaking much more cost-efficient, make the pipeline easier, and be able to immediately see exactly what we have. We built 67 sets in Sydney. We shot the film in 61 days, including five countries. That means we just don't have the time to process negative—we need to strike sets the minute we wrap each day.

How many Panavision/Sony HDW-F900 digital 24p CineAlta cameras were involved altogether?

McCallum: Six. We were in the desert in Tunisia in the middle of summer, where it was easily between 120 and 140 degrees. We had a huge rain set where we shot in the rain for weeks on end. We shot in every different kind of conceivable situation and had backups. And as it turned out, we only shot with two cameras. We never had a single problem with our A and B camera throughout the whole shoot. We only used our C and D camera for second unit.

We had to be diligent about checking the back focus, because the imaging element is sensitive to changes in temperature. But since you can see everything you shoot on an HD monitor, it's totally and readily apparent when focus is off. We had no problems in terms of equipment failure whatsoever.

What role have manufacturers played in making this project possible?

Meyers: Throughout the process, we periodically met with vendors to request features that would allow us to integrate the digital cameras into our established pipeline. Besides just the format, the other issue was the camera package itself, most specifically the lenses. How do you support widescreen? How do you support the capture of images that can fit into that pipeline for the type of work we do? Not just live-action material, but also going into a workstation and dealing with process screens, extractions, repositions, all those things.

Beyond lensing and the format was a whole slew of support issues that had to do with what we're recording on: How do we get the signal out of the camera? How do we display the image? We focused a lot on the way the data are recorded to tape, the look-up tables in the camera. We focused a lot on coming up with a standard, a calibration so that the camera could always be returned to a certain look, a certain response, and knowing exactly what it would be.

In terms of capturing an image, what's most important is knowing how the camera responds to a scene insofar as working with digital images and incorporating them into the scene. We have lots of experience in working with CG characters and lighting 3-D scenes that we have modeled as physical environments. We've had many years to explore the response of film. So the key thing in going digital is knowing the response of those cameras. We very carefully determined how we could modify that, and we came up with ways that, for this current first generation, we could put as much information onto the tape as we could and later extract it to blend those images with the CG.

Who provided the lenses?

Meyers: Panavision. They did the support package too, which is another area of their expertise. So we were able to put cameras on dollies, Steadicams, cranes, scaffolding, and other special rigs with the same accessories and familiarity of controls that the crew had with film.

At what ASA did you effectively rate this camera?

Meyers: We found that we wanted to run the cameras as open as possible, so we were very close to T2, sometimes wider and sometimes just closed down from that. And what we found is that we were right up against a noise floor in the camera. So I operated the camera often, when I had the luxury of doing it, with reduced gain, which improved noise.

That's minus 3 dBs, correct?

Meyers: Yes. What that ended up doing is rating the camera at around ASA 400. When you add an electronic shutter, to simulate a 180-degree film-camera shutter, it knocks that in half. So you're down to around ASA 200. When exposing for the highlights, we're around 320 ASA.

Was David Tatersall, your DP [director of photography], involved in the testing procedure with you?

Meyers: Most definitely.

What challenges did you have in working with that new equipment?

Meyers: The biggest challenge was to integrate all the pieces of equipment, interconnect them in a way that we could set it up on a film stage without impacting the operations of the shoot schedule.

I set up a vision-control area slightly off the set, and worked via radio with camera assistants and had a setup where David could move between the set and this controlled viewing environment to work out the fine-tuning of the setup of the cameras. So he was back and forth between the set and this engineering area, again, without impacting the schedule. From a logistics angle that was huge, as you can imagine when shooting two to three cameras, moving them 36 times a day, and each time having to protect equipment that's interconnected with many cables.

That took quite a bit of doing not only in terms of the configuration of the equipment such that you could recable and move continually, but even before we got to that stage, getting all the other bits and pieces organized: the monitoring, the distribution, the time code, the computer-control systems for converters that are needed to interface with video assist and with production audio.

Were there any changes in the amount and/or quantity of lighting instruments needed for working in HD?

McCallum: It's indistinguishable from shooting with film. Our lighting package on *Episode II* was virtually identical to *Episode I*, and we shot on the equivalent of 10 stages.

HD can't be undercranked like film; was that a concern?

McCallum: Not at all, because we shot pretty traditionally. Most of our high-speed work is in postproduction, and we're working on software solutions for that, which are fantastic. It's the high-speed stuff for pyro work that's our next big challenge.

The HD aspect ratio is 16:9 and Panavision anamorphic is 2.35:1. What aspect ratio was Episode II filmed in and how was this accomplished?

McCallum: We shot in HD native aspect ratio of 1.78:1. This allows us to choose our 2.35, our anamorphic crop within the frame. It's like shooting VistaVision. We can extract our 2.35 frame and crop anywhere we want to in the exposed frame, which for George is a really powerful instrument because he loves to cut and paste one section of a frame and use it in another section. It allows you to pan, reframe, and start from the bottom of the frame and pan up to the top of the frame if you want to pan it. We can also safely blow up any image by a factor of 100 percent. The most we could even get away with on 35mm was 15 percent and the most on VistaVision was 17 percent.

Did you find that the camera's longer load times enhanced the actors' performance?

McCallum: Absolutely. I don't know a single director in the world who hasn't been frustrated by the time limits of film stock. Whether you're doing improvisation or a complex stunt where you don't want to have to stop and reload, digital allows you to keep the camera running.

When you're shooting traditionally, it always takes five minutes to reload a camera by the time you break the rhythm, check for hairs in the gate, reload, and get ready again. With 36 set-ups a day, it can add up to an hour and a half to two hours of just hanging around waiting.

With digital, we reloaded once a day. During our lunch break, an assistant gathered the tape from the camera. We'd already made a simultaneous clone of the master at 23.98 and a down-conversion to Digital Betacam for editorial. One tape goes into a vault as the master. Another is the safety master. The down-converted Digital Betacam tape from the morning then goes straight into the Avid in real-time. By the time the editor comes back from lunch, he's ready to cut. Everything's logged. It's so easy.

We don't have to sync dailies, because we sent the audio feed directly to the camera masters and also to the down-conversions. There's no need for traditional dailies. George, myself, and the DP watched the footage on the set at full resolution. At the end of the day, you can go to the editor and see your work basically assembled. We can screen our dailies right in our theater directly off the Avid, and the quality is fantastic.

Did shooting with 24p HD allow you to do more offline work on location?

McCallum: We had a lot of our artwork, certainly some wire frames for animation, all with us. And as we're making the film, it's very easy to comp hi-def, but it's no different than a video tap off of film. It's just that the quality level allows you to make much better judgments because it's so good.

Also, we shoot a lot of blue screen and green screen, and it's often hard for actors or actresses to come into a totally blue room and understand what they're acting against. So we block out the scene, record that take, and then they can see it with the model or the artwork—even if it's just simple artwork—comped-in. Then the actor gets a sense of the perspective of the room, how big it is, and what's going on. We're watching on two 42-inch Sony plasma monitors, one for each camera; it's like a small theater.

How did postproduction proceed after principal photography?

McCallum: The editorial process is ongoing. Periodically, we send Avid sequences to ILM, they make EDLs, and digitize selects to their server. That then goes into their pipeline and they can also output directly to the TI [Texas Instruments] HD cinema projector for dailies. At this point, we're strictly watching effects shots in HD off the server from the digital projector.

Is the HDCAM 24p 3:1:1 signal being followed all the way through, or is it composited with 4:4:4 material from the server?

McCallum: It depends on what the origination is. For our stage shoots, we're now recording to the RGB 4:4:4 output directly to a hard drive. These are the shoots being done at ILM. The production footage was all HDCAM compression recorded on tape. ILM has some proprietary magic that they work on our material to optimize it for their pipeline. But it's essentially the HDCAM tape that we're using as our masters.

The picture quality is the best we've ever edited with, and we have three terabytes of storage for editorial here at the Ranch. We have every single frame that we shot online because nothing ever goes out of play in a George Lucas film. We can repurpose things that would normally be an NG take. George can say, 'There's an element in there that I want to use elsewhere.'

Can you elaborate on the picture quality of the post-material?

Blanchard: I'm referring to the fact that there's normally so much material involved in a feature film that you end up working in highly compressed Avid resolutions like AVR 3 or AVR 4. But the price of drives continues to drop and network technology keeps improving, which enables us to keep all of this material online with a lower compression rate and higher quality. We're now at a 14-to-1 compression rate.

What kind of a network are you using?

Blanchard: Fibre Channel with Avid Unity. In the past, you couldn't put that much storage on a nonlinear system. You'd have to make huge compromises in the level of resolution you're seeing in the editing process, and that affected the picture because you couldn't really see a lot of details or know what you had. Now we're screening the film into our theater at the Ranch off the hard drives at a quality level that we didn't have until the online stage of *Episode I*.

We understand that you had a software program developed so that images could be played back on the set and viewed as they might be seen in projection?

Meyers: We did a couple things. We wrote our own control software that took control of the Sony camera, the serial interface. And that was run on a laptop. From that, I could load in 'looks' in the camera. So, for example, if I needed to make an adjustment to see how something might be timed later, I'd do that just in controlling the camera and then I could go back and shoot in optimal fashion. We also built some other software that was more for the studio photography. And that loaded curves into a digital-to-analog converter that would then modify the camera image so that when it was displayed on the monitor, it would appear the way we intended it to.

So there were two halves of that. We had software controlling the cameras, which could also load curves into the camera, and then we had curves that we could load into devices that affected a display, a display lookup, if you will. And that was also rippled all the way downstream into the post-process so that the web stations could display that image as it came off the camera, but show it according to how it would end up getting printed to film or digitally projected.

McCallum: Fred was on the set as HD supervisor. We needed David to be able to understand the gamma tables so he'd know the latitude he had on set in terms of exposure; that's really the skill set that a cameraman now has to bring to it. We had a cable that went from the camera to a splitter that went straight to a special monitor for George and a monitor and computer set-up next to the HD recorder that David could see.

While David watched the image on the hi-def monitor, he could pick any frame he wanted, double-click a mouse, and it would grab that frame from the hi-def. He could make his adjustments, check the latitude of each frame, and—most importantly—at the end of the day, from the takes that George had approved, we could download those single frames to ILM. ILM could then actually build "wedges," in other words, build the information that they needed to be able to expose that same image to make it represent itself exactly onto film. When we were shooting in Australia, we used the time difference with California so that ILM e-mailed the wedges back to us the following morning, enabling the DP to know his wedges for every single frame in the film.

It's awesome, because it means that we are effectively grading the movie as we're shooting it, and once we start to make our cut, all that information is in our software program, which allows us to expose the film exactly the way we want to. Talk to anybody who has to go through trying to come up with their first answer print. It's a nightmare.

You're shooting now in postproduction on stages; what cameras are you using?

Meyers: We're using both the HDW-F900, and the newer HDC-950. The 900 is slightly bigger and has the internal camcorder. The 950 requires recording externally, but for studio applications, that's not an issue. The fact that it's slightly smaller and has some fiber interfaces to it allows us to integrate it a little bit more easily into some of our elaborate camera rigs here at ILM.

Was audio recorded on the HDCAM?

Meyers: Yes. It made it much simpler for down-converting in editorial to keep audio a reference track there. That worked out very well. We also recorded it direct to disk and to DAT. One of the advantages of recording the image digitally is that we were able to streamline a lot of the production audio and editorial post-processes by carefully selecting frame rates, time code, gen-locking, and sync signals such that basically material for production audio could go immediately into the tools for audio post without sample-rate conversion or rate conversion. And, similarly, down-conversions of the material could immediately go into editorial without syncing audio or also any rate conversions required.

The video signal path of the material that you recorded on location and throughout the postproduction process was HDCAM, correct?

Meyers: Yes, we recorded both in-camera, and for the purposes of the protection systems, we did a simultaneous safety with an external HDCAM recorder.

And that was through an SDI [serial digital interface] adapter?

Meyers: Yes.

And it was all 4:2:2?

Meyers: What we've done with the principal photography material is bring that into a 4:4:4 base, using our software-conversion tools. And we're treating all the postproduction as if it was in a 4:4:4 RGB space. So we've converted all the material we have. We've built systems that do that on the fly and then write the HDCAM material out to a file system that is in the same format that handles our rendered material, allowing us to just plug right into that pipeline process. What we've done since principal photography is add other capability at ILM with the fact that we're a fairly fibered facility here and we can interface our servers with our stages. We also now have the ability to record uncompressed-to-disk systems both in the 4:2:2 YUV format as well as the 4:4:4 RGB format.

From which camera system or what media?

Meyers: To hard disk. Currently, the first systems were SGI-based workstations that handle HD rates real-time. And we're also looking at some other vendor equipment that might be applicable. And, similarly, that's what we're doing with the 4:4:4, the RGB mode, is go to disk-based recorders that have been basically hot-rodded for that application.

After you create your final digital master, how will you make film copies for distribution to theaters?

McCallum: Remember, our "negative" is on a server. We'll go straight from our server to an IN [internegative], and from the IN we go straight to release prints. So we get to bypass the original negative, the answer print, and the IP [interpositive] stage. We'll be releasing on film from 2,000-foot negative reels without a single splice. The level of quality that we will provide to 5,000 theaters will be unprecedented.

What about releasing for digital projection?

Blanchard: That depends on what technologies are in place by the time we're ready to release this film. We have several plans, but we're going to wait until the last minute to optimize that technology and leverage what will happen in the next few months so that we can show the best possible image in as many theaters as possible.

McCallum: There are a lot of exhibitors who want digital projection, but the industry has to figure out who's going to pay for it. I believe that a consortium will come together with a fiscally responsible business plan that'll actually make it work. However this comes about, it *is* going to happen and nothing will be able to stop it. And the great thing about it is that the average person that goes to see a movie will see a copy that perfectly replicates the 'original negative.'

Meyers: As we finalize shots through our pipeline to the digital projection system here, we noticed that the images are devoid of the gate weave and instability that most people associate with film projection. So far, there have been discussions about adding grain and weave. We may; we may not. I think if the image looks pleasing in a digital format, we would want to keep it that way.

Any closing thoughts on the digital filmmaking experience of Star Wars Episode II?

Meyers: It's an exciting time. We're excited about how well things fell in place. We've got some technology now that will enable us to move forward and we're not going to stop. We'll keep upping the bar and, I think as we pull the whole process together digitally, we'll actually get even more of the benefits of staying digital throughout.

Blanchard: At the end of the day, it's all about enhancing creativity and the story-telling process. That's what drove the whole initiative. George likes to be able to do anything he can think of when he's in the post-process and he wants to be able to spin on a dime and say, 'You know what? I was going in that direction. Now I'm going in this direction.' And, you know, our mandate is just to make it easier for him to do that. He just wants people to be able to actually see these movies the way we see them and the way he intended them to be. And I think we're getting really close to that.

McCallum: Digital technology really gets down to one simple fact. A writer can write anything he wants to now. A director is only limited by his imagination. A producer can't say 'no' anymore, because now there *is* a way to solve each production challenge and to do it in a cost-efficient, fiscally responsible way. It doesn't mean that by using the technology that the film is going to be any better—that's still about talent.

Epilogue

Star Wars Episode II: The Attack of the Clones opened on May 16 and 17, 2002 and grossed more than $310 million in U.S. theaters alone. Any doubts that a digitally photographed motion picture could be profitable were vaporized as if by the Jedi Knights themselves. As this book goes to press, George Lucas and his team are in postproduction on the next installment in the digitally photographed *Star Wars* saga, titled *Episode III: Revenge of the Sith*. The film has been shot using Sony's new improved HDCAM SR format.

Digital Compression

With the motion-picture industry well-established and poised at the dawn of a second century of growth and profitability and filmmakers such as George Lucas determined to usher in a new era of film technology, the mass media began to report on the new age of "digital cinema." Movies could be "photographed" on HD videotape, edited on "nonlinear" Avids, and even projected "digitally" in theaters. Films could be made and shown without film. Digital cinema held the potential for providing filmmakers with improved and—possibly—more affordable creative tools. But what exactly is this thing called "digital?"

As noted in Chapter 2, it takes millions of 0s and 1s to digitally record and store even a few moments of motion-picture footage. Digital image creation is a product of the computer age, and the computer chips and hard drives that process and store movie-image data can do their job even more efficiently if some of those 0s and 1s can actually be discarded in a process known as *compression*. When done correctly, this process causes no discernable loss of picture or sound information. It is useful, then, in understanding digital cinema to also understand what digital compression is. This chapter is the most technical of all in this book, but don't despair. Understanding digital compression is not mandatory for filmmakers and people interested in motion-picture technology. After all, you don't need to be a mechanic to drive a car, and you don't have to be an electrical engineer to be a filmmaker. Today's digital cinema cameras, recorders, and postproduction systems are made to be used by creative talents, not scientists. There are those creative talents, however, who would argue that a solid base of technical understanding makes them better at what they do. For them, we present this chapter.

Managing Digits

Digital compression is a term used to describe technologies that reduce the quantity of data that—in the case of digital cinema—represents motion-picture footage. In general, compression is a solution to a problem, so before describing the solution, it makes sense to identify the problem.

Traditionally, movies have been delivered to theaters as 35mm release prints. These prints carry both a sequence of images and a variety of analog and digital sound tracks for different audio systems (see Chapter 8). In terms of film, the images are just that—little transparent photographs that you can view just by holding the film up to a light. To create the illusion of a moving image there are 24 images for each second, which comes out to more than 170,000 for a two-hour movie. Analog sound tracks are coded by density or area. Digital sound tracks are compressed and then coded in a way that permits reliable recovery of the data in the projector system. Strangely, it seems probable that digital cinema will see an abandonment of compression for audio; although compression will be used for the image data, current thinking leaves the audio data uncompressed!

When a movie is digitally played back and projected in a theater, the release print is replaced by a file of digital data that may be distributed by a physical medium such as a DVD, or by transmission over fiber or a satellite system. Obviously, the data must represent all of the images and sound necessary for a high-quality full-motion theater presentation identical to what you would see projected off of 35mm film.

Uncompressed audio represents quite large quantities of data. Usually digital audio is coded at 48 kHz, and at least 16 bits per sample. So, for a two-hour movie, each audio channel is nearly 700 megabytes of data (equivalent to approximately one compact disc). For a movie with, say, eight sound channels, the audio data alone would come close to filling a complete two-layer DVD. That being said, however, compared with the data necessary to represent the movies images, the audio data are almost insignificant. The amount of information in a 35mm frame is stunning. The lowest quality being considered for mainstream digital projection is known as "2K," which means that each image is represented by an array of about 2,000 pixels from left to right. The number of rows of pixels depends on the *aspect ratio* (the relationship of the width of the picture to its height) of the movie. For an aspect ratio of 1.78:1, the most likely figures are 1,998 x 1,080, for a total of 2,157,840 pixels for each image. Most experts would say that we need at least 10-bit accuracy for each of three color components (usually red, green, and blue). So each pixel would take 30 bits, or 33/4 bytes, or about 8 megabytes for each image. At this resolution a two-hour movie would need almost 1.4 terabytes.

The architects of digital cinema are setting their sights higher than this. The current draft of the DCI specification suggests "4K" resolution, and 12 bits per component, using X, Y, Z coding (DCI refers to Digital Cinema Initiatives, an organization formed by seven Hollywood studios to propose standards for digital cinema). This translates to nearly nine million pixels per frame, and over 6 terabytes for a two-hour movie, which means nearly 12 terabytes for a three-and-a-half-hour epic!

What does this mean in practice? Even 6 terabytes would need more than 700 dual-layer DVDs (plus another for the sound!). A satellite transponder typically provides a data rate of about 27 megabits per second, or a little over 3 megabytes a second. Even if we assume perfect transmission with no allowance for error correction, transmitting our epic would take well over a month.

Clearly, this is totally impractical. Digital cinema cannot realistically happen unless we can substantially reduce the amount of data needed to deliver the movie to the theater. This is the role of compression.

Lossless and Lossy Compression Systems

Some compression technologies are lossless or perfect—the output is guaranteed to be identical, bit-for-bit, with the input. These systems generally do not provide enough compression to be useful for moving-image applications, but they do have a role to play that will be discussed later.

Most compression systems are lossy or imperfect; the final output is not identical to the input. In other words, compression systems introduce impairments. Some of these are quite familiar today, and include softness, smearing, pixilation or blockiness, banding, or visual defects on sharp edges, etc.

Expectations

The choice of compression system should be made, if possible, so that the impairments introduced are insignificant in the context of the application.

Some applications, such as low-bandwidth video streaming to personal computers, can tolerate significant impairment. Others, such as cable or satellite distribution of broadcast video, need to provide a quality of experience comparable to conventional television.

Digital cinema, however, is a very demanding application. A reel of 35mm film typically provides superb image quality, so if a digital cinema file is going to replace it, it will never be accepted for delivery if its picture quality looks inferior.

In reality, however, most 35mm presentations fall far short of the ideal capabilities of the medium. Scratched prints, projector lamps that aren't bright enough, out-of-focus projectors, and all the other sloppy characteristics of far too many movie theaters all contribute to an average viewing experience that is well below what can be achieved. Outside of the best screening rooms and theaters in Hollywood, New York, or other major cities, movies shown on 35mm film often lack the picture quality they should ideally have. That being said, for digital cinema to even begin to gain acceptance (which it has), it will have to match or exceed the best of 35mm projection—even though this "best" is rarely seen. Digital cinema frequently will be offered as a premium product, and audiences will be led to expect an experience that improves on conventional cinema.

So, digital cinema presentations must be free of the artifacts viewers have come to expect from 35mm. Even more important, digital cinema must not introduce new picture impairments, such as those associated with compression systems. This has led to the concept of *visually lossless* compression. Even though a compression system for digital cinema cannot be totally lossless, it is required that the output be indistinguishable from the input under real theatrical viewing conditions.

How long such requirements will survive is uncertain. More bits cost more money, and it may be that economics will force acceptance of lower-quality standards. Nevertheless, it is a basic requirement that the digital cinema system be able to prove this degree of performance, even if commercial realities force compromises down the road.

Principles of Compression

As discussed above, there are two fundamentally different types of compression. *Lossless compression* is the function offered by such computer programs as PKZip and Stuffit that offer perfect reconstruction of the original. These systems work by finding the most efficient way to code the data.

Unfortunately, lossless techniques alone do not provide sufficient compression for applications such as digital cinema. To reduce the image data substantially, *lossy compression systems* must be used. Lossless tools, however, still have a role to play. Because lossless systems guarantee perfect reconstruction, they can be used on any data, including the output of a lossy compression system. So, a lossless compressor is usually the last processing block in a lossy compression system.

Lossy compression systems do not provide perfect reconstruction of the original data, but they use a number of techniques to ensure that the gain from the compression is much greater than the loss in fidelity. In the case of image compression, the goal is to maximize the savings in data required to represent the images, while minimizing visible changes to the images.

It is possible to do this because the human visual system (HVS) has limitations. People are typically unaware of these limitations, which actually works in favor of compression. The HVS can operate over enormous contrast ranges, and it can detect very fine detail, but it is very insensitive to fine detail (known technically as "high spatial frequency") at low contrast. The result is that fine detail at very low contrast can be eliminated completely without visible change to the image. Furthermore, even where the fine-detail low-contrast information cannot be completely eliminated, it can be transmitted at much lower precision without affecting the viewing experience.

In the following sections, we will examine the process in more detail, but quite simply, this is *image compression*.

Process of Compression—Generating the Data-Sampling and Quantization

This is not strictly part of the compression process, but understanding the data that need to be compressed is essential to understanding how the compression works.

Digital data are generated by sampling images. These can be done directly in an electronic sensor in a camera, or by scanning a film that has already captured the images. Sampling measures the amount of light that falls on one small area of the sensor (known as a *pixel*). This measurement produces an analog value—in other words, within the range of measurement, there are no restrictions on values. For these values to be useable in a digital system, they must be quantized, or approximated to a number of preset values that can be represented by a given number of bits.

Let's take an example. To make it easy, we'll suppose that the values from the sensor are very convenient; black is zero and the brightest white is 255. When we sample, we may get any values such as 3.1, 48.79, 211.3, 72.489, etc. In this example, however, we are going to use an 8-bit digital system, and an 8-bit value can represent any of the integers from zero to 255, but nothing in between. To quantize the measured values, we have to replace each value by an integer, usually the nearest integer. So the examples above become 3, 49, 211, and 72, and every quantized value can be represented by an 8-bit word.

It is important to recognize that the quantization process introduces errors, known as *quantization noise*, because each measured value is approximated to the nearest permitted value. The number of bits used is critical. As stated, 8-bit quantization permits 256 values. Higher precision may be used; 10-bit quantization provides 1,024 values, 12-bit quantization—as proposed for digital cinema—provides 4,096 possible values. More bits representing each sample means more total data, but not necessarily more bits after compression. Noise, including quantization noise, is difficult or impossible to compress, so the more accurate representation may compress better even though it starts off with more bits.

Color Space Conversion

In the previous section, we discussed the sampling and quantization of a signal. Having only one signal, however, provides only monochrome information. Image sensors for color need to sample three different color ranges, and generate three signals, generally known as red, green, and blue or "RGB."

We could compress each of the R, G, and B separately, but it may be advantageous to change to some other representation. Brightness changes in a scene affect all three signals, so some information is repeated in all three channels. It is possible to change to other representations that have less redundancy (mathematically we say "less correlated"). The most common alternative is Y, R-Y, B-Y (also expressed in the digital world as Y, CB, CR)—the luma and color difference signals used in video. An alternative has been suggested as Y, CO, CG (luma, offset orange, and offset green). These color difference components provide similar benefits, but can be implemented using integer arithmetic.

At the time of this writing, the DCI draft specification calls for conversion of RGB to XYZ color space, and for independent compression of the X, Y, and Z signals. There may be room for improved efficiency, but more research is needed.

Transforms

Strictly speaking, color-space conversion is a *transform*, but when we speak of transforms in the context of an image compression system, we generally refer to some mechanism that converts spatial data to spatial frequency data.

It was stated above that the HVS is very insensitive to high-frequency information at low contrast. Unfortunately, if we look at an array of pixel-brightness values, it is very difficult to extract frequency information. The transforms used in image compression systems convert spatial brightness data into data that represent the spatial frequency components of the information.

This is done by defining a number of basis functions, each representing spatial frequencies that may be present in the image. The basis functions are chosen so that some combination of multiples of these functions-when added together—can represent any possible array of pixels.

The set of basis functions for the Discrete Cosine Transform (DCT)—a transform much used in compression systems—is shown in Figure 4-1. Each basis function is itself a block of 64 values. This type of compression system divides the image into square blocks of 64 pixels, and applies the DCT to each block. This generates a set of 64 coefficient values, each representing the quantity of one of the basis functions required to construct the original set of 64 pixels.

As a very simple example, in the top left basis function (known as 0,0), all 64 values are the same. If the original pixels were all the same (a flat area of the image), the transform would be just a multiple of the 0,0 basis

Figure 4-1

The set of basic functions for the Discrete Cosine Transform.

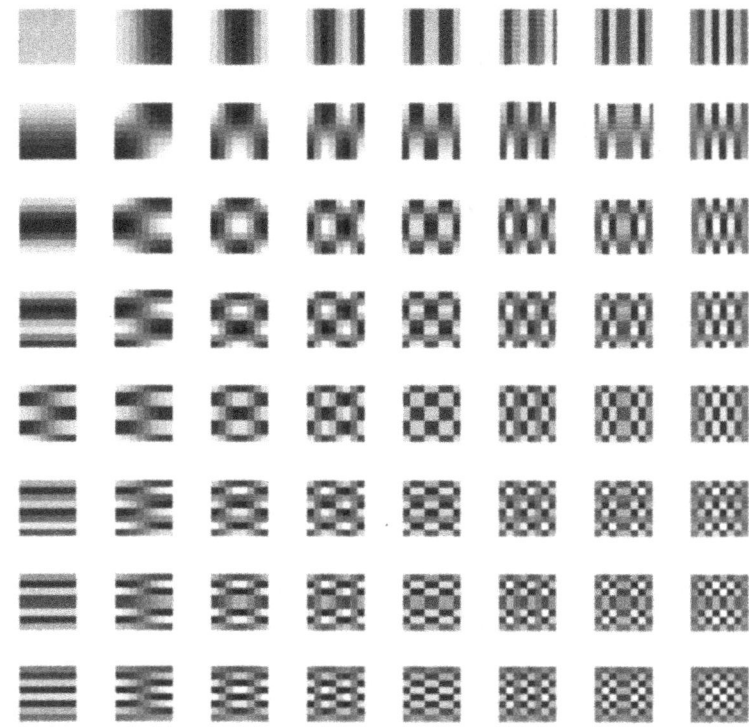

function, so the coefficient for 0,0 would be nonzero; all the other 63 coefficients would be zero.

DCT is used in most of the compression systems standardized by the Moving Picture Experts Group (MPEG), and is the dominant technology for image compression. In particular, it is the core technology of MPEG-2, the system used for DVDs, digital television broadcasting, and that has been used for many of the trials of digital cinema.

A variant of DCT—using only integer arithmetic—is used in the latest MPEG scheme, the MPEG AVC (Advanced Video Codec), also known as MPEG-4 Part 10, or ITU-T Recommendation H.264, or just JVT (for the Joint Video Team that developed it).

The other transform used extensively for image compression is the wavelet. A typical wavelet in analog form is illustrated in Figure 4-2, which shows a one-dimensional wavelet. Two-dimensional wavelets are possible, but in image work, they are generally implemented by applying horizontal and vertical one-dimensional wavelets separately. A wavelet is a burst of energy that can be used to extract a band of frequencies from an image. Unlike DCT, wavelet systems do not generally divide the image into blocks, but the wavelet is moved, pixel by pixel, row by row, across the entire image. In a practical application, compression is performed in the digital domain, so the wavelet is represented by a sequence of sample values, as shown in Figure 4-3.

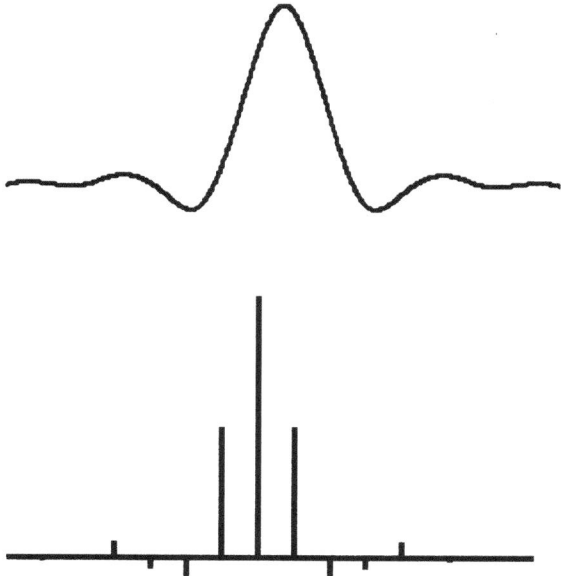

Figure 4-2

A typical wavelet in analog form.

Figure 4-3

Wavelet, as represented by a sequence of sample values.

The wavelet can be used to separate the highest band of spatial frequencies from an image. When this has been done, the remaining image has no high-frequency content, so fewer pixels are needed; in fact, the pixel density may be halved in each direction, leaving a new image with only one quarter of the pixels. Because of this, the same wavelet may be used again to extract the next band of frequencies—and so on until there are only a few pixels left! Wavelet systems have only a single basis function.

Wavelets have been used in a number of systems, but the technology is more processor-intensive than DCT, and it has yet to see widespread deployment. The wavelet transform is the core technology of the JPEG2000 image compression standard, and this system has been selected by DCI for the "official" deployment of digital cinema.

Like DCT, a wavelet transform produces coefficients representing the spatial frequencies in the image. Neither transform in itself reduces the amount of data representing the image; in fact, the data increases somewhat through this step. The data are, however, now in a form more suitable for exploitation of the limitations of the HVS.

Quantization

We discussed quantization at the beginning of this section, in the context of the precision that would be used for representing the original pixel data. Now that the data have been transformed into the frequency domain, we can reassess the need for precision.

When natural image data are transformed, an enormous number of coefficients is produced but—as a practical matter—a very large proportion of these coefficients are zero, or very close to zero. Setting the near-zero coefficients to exactly zero has virtually no effect on the image when it is transformed back into the pixel domain.

It was stated earlier that the HVS is very insensitive to low-contrast fine detail. Now that the data are in the frequency domain, we can recognize that content; it is represented by all the high-frequency coefficients that have low amplitude. These can also be set to zero without impacting the viewing experience.

It is also found that the HVS is limited in a slightly different way. Even when the fine detail has sufficient contrast to be significant, the actual magnitude can be represented very coarsely without affecting the appearance of the image. The pixel representation may require, say, 10-bit precision, but the highest-frequency coefficients may need only 2-bit precision (just four values). In other words, the HVS is satisfied by indicating that there is zero, a little, a moderate amount, or a lot, of that spatial frequency.

These three steps can be accomplished by a single processing stage that requantizes the frequency-domain data, using increasingly coarse quantization for higher spatial frequencies. This forces the near-zero coefficients to zero, and represents the high spatial frequencies with fewer bits than the low spatial frequencies.

This process is similar for any of the transforms discussed, and is the (only) point in the process where loss is introduced. It is also the (only) point where the degree of compression can be controlled, by adjusting the settings of the quantizer. The result of the quantization process is a large number of frequency-domain coefficients, but many of these are zero, and many more are represented by only a small number of bits. In itself, this still does not help a great deal for transmission, but these data are now very suitable for lossless compression.

Lossless Compression

Lossless compression guarantees that after decompression, the result will be identical, bit for bit, with the input. So, after transformation and quantization, we can apply lossless techniques without introducing any further impairment.

Lossless compression systems reduce the amount of data by using the most efficient coding schemes for the data. This may sound simple, but this type of compression is a science in itself, and we can only look briefly at a few of the techniques that are used.

The simplest is *run-length encoding*. If the data contain long runs of identical values, it is easy to arrange a special code that tells the decoder that, for example, (code) 46,23 should be interpreted as the value 46, 23 times. Clearly, with suitable data, this can provide substantial coding gain.

Prediction also relies on known characteristics of the data. Even in a bitstream from a lossy compressor, it is often the case that the presence of certain values at one point carries a high probability that the same, or similar, values will appear at some other point. Prediction schemes use this probability to reduce the number of bits to be transmitted when the prediction works, and accept a penalty on the (hopefully much fewer) occasions where it does not.

Variable-length coding uses the fact that generally some values—or sequences of values—occur more frequently than others. It is advantageous to use short codes to represent the frequent values, at the expense of longer codes for the less frequent values. A simple example of this is Morse code, which uses a single dot for the letter "e" (the most common letter in the English language) and much longer codes for less common letters such as "q" and "z."

Arithmetic coding is a system developed by IBM, originally for fax machines, which uses continual analysis of the bitstream statistics to generate an efficiently coded output.

Modern compression systems use these techniques and improved variants that adapt themselves to changing characteristics of the bitstream, such as content adaptive variable length coding (CAVLC) and content adaptive binary arithmetic coding (CABAC).

All lossless compression relies heavily on statistical techniques and, if there is a good understanding of the data, it is possible to arrange the data in ways that help the process. For example, DCT systems produce blocks of coefficients containing many zeroes. The blocks are scanned in a zigzag manner that tends to maximize the run length of zeroes in the stream that is sent to the lossless compressor.

Temporal Compression

Some compression schemes, particularly those developed by MPEG, can take advantage of additional redundancy in a moving-image sequence. A "moving image" is really a sequence of still images, and in a normal scene, consecutive images are very similar. It is possible to predict content of one image by identifying similar image fragments in one or more frames that have already been transmitted. Provided the decoder has been instructed to store the earlier images(s), it can be instructed to reuse parts to form the current image, saving considerable data in transmission.

This is another complex topic, and as temporal compression is not currently part of the DCI recommendation for digital cinema, it will not be discussed further in this chapter.

Caveats

It is very important to remember that all compression schemes are designed based on assumptions about the information being coded and the way in which it will be used. Particularly important for image compression schemes are the assumptions made about viewing conditions, especially viewing distance. The contrast sensitivity function of the HVS relates to perceived spatial frequency, measured in cycles per degree. These assumptions work only for a known minimum viewing distance. If the viewer goes closer to the screen than assumed in the compression scheme, the assumptions will fail. If the viewing distance is halved, the apparent spatial frequency is (approximately) halved!

The other major area of danger is in implementation in a commercial environment. For purposes of digital cinema, any of the principal compression technologies is capable of providing a "visually lossless" experience if enough bits are allowed in the compressed bitstream. Equally, all the schemes will fail if commercial constraints dictate excessive compression.

Compression Schemes for Digital Cinema

At the time of this writing, there has been only small-scale experimental deployment of digital cinema. These deployments have mostly used MPEG-2 (as used by DVD) or a proprietary wavelet-based system. As the time approaches for digital cinema to become far more mainstream, there is a strong desire to select and standardize a single system that provides the required transparency and that is not burdened with onerous licensing conditions.

DCI has selected JPEG2000, currently believed to be royalty-free. It has been shown to be capable of the performance required and has one big advantage. As described above, wavelet schemes encode successively smaller pictures. This means that contained in a bitstream representing one resolution is the data for lower resolution codings of the same images. Specifically for digital cinema, a stream that is encoded at 4K resolution inherently contains the coding of the 2K stream. The decoder for a 2K projector can decode the 2K data, and ignore the rest of the bitstream.

JPEG2000 is computationally very intensive, so this decision is not universally popular. If problems arise with implementation or licensing of JPEG2000, the most promising alternative is probably the fidelity range extensions (FRExt) of the MPEG AVC that was standardized in 2004. The movie industry, however, is very wary of licensing terms that have been associated with recent MPEG standards, so there is a strong desire to make JPEG2000 work for digital cinema.

And that's more than you'll ever need to know about compression. Keep in mind that, as computer hardware and software continues to advance, the capabilities of compression will improve and picture quality will exceed the best that film has to offer. Processing compressed digital image data—whether gathering, manipulating, transmitting, or displaying it—is the underlying science of digital cinema.

Cinematography Tools

The new technologies of the Digital Cinema Age have brought forth more alternatives for capturing moving images than ever before. Film, meanwhile, reigns supreme for motion-picture cinematography as it has since the dawn of moving pictures. Where the evolution of both forms of media will go in the future remains to be seen, but for the present, it's important to remember that film is as much a part of the digital cinema revolution as any HD or SD video format—if not more so. The rise of the digital intermediate (DI) in postproduction (see Chapters 7 and 8) has reconfirmed film's role as an image-acquisition medium of limitless artistic potential far into the future. Today's DI systems enable moviemakers to enjoy the best of both worlds: the beauty and versatility of shooting on film *and* the vast creative tools and artistic control that digital offers.

Of course, the word *film* doesn't refer to a single imaging format any more than "digital" does. Are we talking 35mm or 16mm? 65mm or Super-16? Four-perf or three-perf? Kodak or Fuji? And what particular film stock? There are nearly as many choices in film as there are in digital—or analog video, for that matter. What's important are the esthetic needs of the filmmaker; what kind of image is he or she intending to create? If painters have the ability to choose a wide variety of brushes in addition to deciding whether to use watercolors, oils, acrylics, etc., why shouldn't filmmakers have similarly varied choices? There are some high-profile filmmakers who have said they'll never use film again. And there are those who have said they will "shoot film until the very last processing lab closes down." But more common are filmmakers who know they can use whatever tool it is they need—digital or photochemical—to achieve their artistic vision.

Ten Reasons Why Film Will Never Die

We've all read them at one time or another: newspaper articles proclaiming that "film is dead." The writers of these articles typically enthuse about how a particular film was produced digitally (fair enough), but then jump to the conclusion that this *must* mean the end of film. This occurs despite the fact that every movie advertised in the same newspaper the article appeared in was photographed and released on 35mm.

It's unfortunate, but today's mass media increasingly seems to prefer hitting readers with dramatic, attention-getting "bytes" of simplistic exaggeration instead of well-researched reports from multiple perspectives. *The New York Times Magazine* indicated on March 14, 2004 that Kerry Conran, Director/Writer of *Sky Captain and the World of Tomorrow* (2004), had made that entire movie "inside his home computer." Further on, however, the article reveals that the budget for the picture was $70 million, and that it required three large rooms of computers as well as the many creative talents to produce images on them.

Claims that "film is dead" or that entire movies can be created on one PC only serve to confuse the general public about new technologies. It also does a disservice to the principal medium by which the moving image has been recorded, stored, and conveyed since its inception: film. A "film is dead" mentality is quite simply incorrect. Although increasing numbers of movies *are* being photographed digitally and can look just as good as film, this doesn't mean the *end* of film. Digital doesn't mean the death of film any more than television meant the death of radio.

Here are 10 reasons why film will never die:

1. Film works; it's a proven, mature medium. Film equipment is widely available, adaptable, lightweight, unencumbered by cables or computers, and it functions reliably in every environment. Accessories are enormously varied.

2. Despite differences in aspect ratios and audio formats, 35mm film is the only worldwide motion-imaging standard.

3. There's more than a century's worth of film content in the world's archives, a vital part of our modern cultural heritage. Transferring all of those archives to an electronic medium is cost-prohibitive. And why bother, when you can convert what you need to the video format *du jour*? Those archives will ensure film's viability far into the future.

4. Video formats can become obsolete in just a few years. The 35mm film standard is more than a century old.

5. Motion-picture film chemistry continues to improve (e.g., Kodak's new Vision2 stocks).

6. Digital techniques for transferring (and restoring) film also continue to improve, and as they do, film yields ever-more picture information—detail we never knew was there.

7. The archival life of today's film stocks is at least a century. Far less is known about the stability of digital tape and other storage media.

8. Digital film scanning, intermediate, color-correction, and film-recording options are enhancing film's viability.

9. Film has more definition than current digital HD. Film has a greater dynamic range. Film cameras are cheap. Processed film is human-readable. Film offers color with neither prism color separators nor color filters, both of which reduce sensitivity and the latter of which reduces resolution and introduces aliases.

10. And film is actually a digital medium (grains are either exposed or not). The grain structure is random, so there are no sensor-site alias issues. Furthermore, the grain structure is different from frame to frame, so there is no possibility of a defective pixel or thermal-noise pattern. And the transfer characteristic of film is part of its desirable "film look," the same look that 24p HD emulates.

And one more thing: Film's viability doesn't mean the death of digital, either. What it does mean is there are more and better tools for making moving images than ever before.

A List of Choices

So now that we've made it clear that film is a full partner in the digital cinema revolution, let's take a look at the different digital cinematography alternatives that are available. There are more arriving every year. No format is "better" than another. They are all different tools for achieving the look a filmmaker wants. Some are expensive and designed to emulate film-centric production workflows; others are affordable and accessible to every indy filmmaker. All exist because they work very well.

Any video camera can be used for digital cinema; there really are no rules. A filmmaker's individual aesthetic sense is the only criteria. Camera options can be divided into four categories: digital cinema cameras intended to replace film as an acquisition media, high-definition (HD) video cameras, standard-definition (SD) progressive-scan cameras, and SD interlaced (NTSC and PAL) cameras.

Digital Cinema Cameras

Arri's D20 camera, captures images using a single CMOS sensor that has an active picture area equivalent to 35mm film, so consequently, any 35mm PL-mount lens will work on this camera (Figure 5-1). Arri's D20 uses the same rotating mirror shutter and optical viewfinder designs found in its film cameras. HD lenses—designed for three CCD prism-based imagers—will not work on the D20.

To capture color images using a monochrome CMOS sensor, the light from the lens passes through red, green, and blue filters arranged in a *Bayer pattern mask* before striking the pixels on the CMOS sensor. Bayer pattern filtering follows a set arrangement and each of the six million pixels on Arri's CMOS has an assigned red, green, or blue value. Color

Figure 5-1

Leading film-camera manufacturer ARRI enters the realm of digital image acquisition with its revolutionary D20 camera.

images are recreated from the monochrome data the sensor generates. The first row of pixels alternates between green and red color filters. The second row alternates between blue and green filters. The first row pattern repeats in the third row and the second row pattern repeats in the fourth row, and so on until the entire sensor is mapped for red, green, and blue.

Row 1: G R G R G R G R G R . . .

Row 2: B G B G B G B G B G . . .

Row 3: G R G R G R G R G R . . .

Row 4: B G B G B G B G B G . . .

When you look at a Bayer pattern, you'll notice that four green and four blue pixels surround each red pixel. This arrangement takes advantage of the overlap that exists between the color spectra of the filters. That overlap makes it possible to interpolate accurate color values based on a pixel's position and the color values generated by adjacent pixels. The actual pixel count on the CMOS sensor is 2880 by 2160. The Bayer pattern reduces the number of pixels for each color to 1440 by 1080, although Arri claims higher color resolutions are achieved using their interpolation algorithms.

The data generated by the CMOS sensor is processed using 32 12-bit analog-to-digital converters in parallel, and output in real-time as a down-sampled 1920 by 1080 YUV (4:2:2) HD signal via HD-SDI. Raw Bayer data can be output via dual-link HD-SDI to a disk array for nonreal-time rendering to 1920 by 1080 RGB (4:4:4) images.

Dalsa's Origin is another single-sensor camera design using Bayer-pattern filtration to capture color images (Figure 5-2). Dalsa's sensor is an 8.4 million-pixel frame-transfer CCD. It has an active picture area of 4046 by 2048, which is larger than 35mm film. The Origin camera has an optical, through-the-lens, viewfinder, uses 35mm PL-mount lenses, and is capable of outputting a 4K data signal. At the moment, there are no recording devices capable of handling its 402 (megabits per second (Mbps) data rate nor any with the terabytes of storage required for the Origin's 4K output.

The 4K data workflow Dalsa is envisioning delivers the camera output over four fiber optic channels using the Infiniband protocol developed by Silicon Graphics. The frame size is 4046 by 2048 (that's a 1.98:1 aspect ratio) and the data are in a proprietary interpolated 4:4:4 RGB linear file format that stores 16 bits per pixel and 16 bits per channel. There are reduced 4K bandwidth outputs available: a 12-bit log format, which lowers

Figure 5-2

Dalsa's Origin camera uses a single 8.4 million-pixel frame-transfer CCD; the camera can output a 4K digital signal.

the data rate to 302 Mbps; and a 10-bit log option, which lowers the data rate to 252 Mbps. The camera's native file format can be converted afterwards—using lossless processing—to other file formats such as Cineon, DPX, or TIFF.

In the realm of possible recording options, the Origin offers what Dalsa calls a Super 2K mode. This is a 2048 by 1080-sized image, stored as 4:4:4 RGB, in a 10-bit log file format, and output via dual-link HD-SDI. It should be possible to record this under-200 Mbps signal on a hard-drive array system such as the Director's Friend or on Baytech's CineRam solid state recorder. The Origin also has two monitor outputs, which provide 1920 by 1080 10-bit 4:2:2 Y Cr Cb signals that could also be recorded on tape formats such as Sony's HDCAM and HDCAM-SR or Panasonic's D5-HD.

Kinetta's Digital Cinema camcorder—seen in working prototypes—is scheduled to arrive in early 2005, as this book goes to press (Figure 5-3). Company founder and filmmaker Jeff Kreines designed this camera to suit his *cinema verité* shooting style. The electronics to support what the company calls a *sensor-agnostic* camera are being developed by Martin Snashall, an industry veteran. The Kinetta is a hand-held camera about the size of a 16mm film camera. It uses 16mm PL-mount or C-mount lenses and supports interchangeable sensors—CCD or CMOS designs with a maximum pixel count of 16 million. The first camera prototypes will use

Figure 5-3

Kinetta Founder Jeff Kreines's with his Digital Cinema camcorder, seen as a working prototype at NAB 2004.
(*Photo by Mark Forman.*)

CMOS sensors from Altasens. These sensors have a picture area of 1936 by 1086 and can capture either 1920 by 1080 interlaced or 1280 by 720 progressive images. The raw data from the CMOS sensor will be stored on a removable hard drive array that mounts on the camera in a manner similar to a film magazine. Each electronic storage magazine holds 110 minutes of footage. The power specifications suggest a single 80-watt battery will run the camera for several hours. Kinetta's proposed breakout box, tethered to the camera via a single fiber optic cable, will provide dual-link HD-SDI ports to output a 1920 by 1080 4:4:4 RGB signal or a single-link HD-SDI port to output a 4:2:2 Y Cr Cb signal. The box also has an SDI port so operators can output a down-converted SD video signal.

Panavision's Genesis camera uses a 12.4 million pixel CCD sensor with an active picture area matching the 35mm film format, which allows any of Panavision's existing lenses to be used (Figure 5-4). The sensor uses a proprietary striped RGB filter pattern instead of a Bayer pattern arrangement. It has twice the pixel density necessary to capture images. The extra density allows the image to oversample to reduce aliasing. The camera outputs a 1920 by 1080 4:4:4 RGB 10-bit log signal via dual-link HD-SDI. It also has an HD-SDI monitor output (4:2:2), two electronic viewfinder outputs, a fiber optic camera adaptor, and integrated lens control. Remote camera control options include Panavision's RDC or Sony's MSU or RMB series controllers. The camera's weight and form factor are similar to existing Panavision 35mm cameras; the Genesis camera body

Figure 5-4

Panavision's digital Genesis camera's weight and form factor are similar to its existing 35mm cameras.

weighs 13.25 lbs. and 25 lbs. when the SRW-1 dockable recorder—manufactured by Sony—is attached. Other notable features include support for variable frame rates from 1 to 50 frames per second and compatibility with Panavision's film equipment.

Thomson's Viper camera relies on a prism-based imager and three 2/3-inch CCD sensors, as do most professional HD or SD cameras (Figure 5-5). It has a mechanical shutter, adjustable from 90 to 359 degrees, and an electronic viewfinder. The Viper uses HD lenses in B4 mounts. Thomson's 9.2 million pixel Frame Transfer CCDs are able to capture images at the full resolution of the CCD in either the 1.78:1 or 2.35:1 aspect ratio by remapping the 1920 by 4320 pixel array using what Thomson calls *dynamic pixel management*. Viper departs from traditional HD cameras because operators can decide to bypass the digital signal processing in the camera and output the raw data from the CCDs as 1920 x 1080 4:4:4 RGB, 10-bit log files via dual-link HD-SDI, when the camera is in "FilmStream" mode.

Viper has three other output modes: as 4:4:4 RGB output as full-bandwidth, full-resolution video processed through the DSP for color balance, color gamut, gamma, knee gamma, toe gamma, and detail enhancement; in HDStream, a 4:2:2 HD unprocessed mode, with the exception of white balance, that is similar to FilmStream with lower bandwidth requirements; and multiple HD (YUV) formats. The HD formats that can be output

Figure 5-5

The Thomson Viper FilmStream camera, as seen at its NAB 2003 introduction.

include: 1080P at frame rates of 23.98, 24, 25, 29.97, or 30 fps; 1080i at frame rates of 50, 59.94, or 60 fps; and 720p at frame rates of 23.98, 24, 25, 29.97, 30, 50, and 60 fps. The HD or HDStream modes can be recorded on Panasonic D5-HD, Sony HDCAM, or Panasonic DVCPRO-HD tape. The FilmStream mode can be recorded uncompressed on Thomson's RAM recorder, Baytech's CineRAM solid state recorder, hard drive array systems such as Director's Friend, or S.two's Digital Film Recorder D.Mag system, or compressed using MPEG-4 with a Sony HDCAM-SR recorder (Figure 5-6).

Sony's HDC-F950 camera also relies on a prism-based imager with three 2/3-inch CCD sensors (Figure 5-7). This camera has an electronic shutter, electronic viewfinder, and uses HD lenses in B4 mounts. Sony's 2.2 million pixel FIT CCDs have an active picture area of 1920 by 1080. The camera outputs data as 1920 by 1080 4:4:4 RGB, 10-bit log files via a single fiber optic cable to Sony's SRW-1/SRPC-1 combo or via dual-link HD-SDI to Sony's SRW-1 recorder (Figure 5-8). Output from the HDC-F950 can also be recorded on a SRW-5000 HDCAM-SR recorder or a hard drive array. Multiple HD (Y Cr Cb) formats are supported including: 1080P at frame rates of 24, 25, or 30 fps; and 1080i at frame rates of 50 or 60 fps.

Figure 5-6

The S.Two's Digital Film Recorder D.Mag system for digital cinema recording.

Figure 5-7

Sony's HDC-F950 CineAlta HDCAM camera is the choice of many high-end digital filmmakers.

Figure 5-8

Sony's SRW1 dockable
HD recorder.

High-Definition Video Cameras

The highest resolution HD signal, as defined by the Advanced Television Standards Committee (ATSC), is the 1080-line format, which is 1920 by 1080 with an aspect ratio of 16:9 (1.778:1). Images can be interlaced or progressively scanned. Most equipment designated as recording in the 1080 format actually only captures 1440 x 1080 due to subsampling of the signal. Cameras or VTRs recording 1080 interlaced signals use the letter *i* after the frame rate to designate interlaced recording. Differences between PAL and NTSC frame rates are maintained when recording a 1080 interlaced signal. PAL countries use 1080/50i (25 fps) and NTSC countries, 1080/60i (30fps). Cameras or VTRs capable of recording progressive images in the 1080 format are indicated with a *P* or *PsF* after the frame rate.

PsF is an acronym for *progressive segmented frame recording*, developed by Sony to record progressively scanned images on HDCAM recorders, which are interlaced recorders. Each progressive frame is split into two fields in order to record it on tape. Because each frame is captured at a single temporal moment, there's no vertical movement between the fields as there is when interlaced scanning (as in NTSC and PAL video) is used. Consequently, there's no loss of vertical resolution due to the interlace factor when the image is reassembled for display or postproduction.

Another HD signal defined by the ATSC is the 720p format. The specification calls for an image format of 1280 by 720 with an aspect ratio of 16:9 (1.778:1). The equipment manufacturers decided to make this HD standard a progressive format, recorded at 60 frames per second, to more closely match the vertical resolution of the 1080i format. Camcorders recording the 720p signal in the DVCPRO-HD tape format subsample the signal reducing the image format to 960 by 720.

Panasonic's Varicam camcorder has a prism-based imager with three 2/3-inch one million pixel IT CCD sensors that capture 1280 by 720 progressive images at 60 frames per second (Figure 5-9). These images are subsampled to 960 by 720 for recording on the DVCPRO-HD tape format. The Varicam camera can alter the rate it captures the images, so the frame rates can be varied from 4 to 60 fps. The recorder section records these frame rates at a constant 60 frames per second. What is recorded in each of those 60 frames varies based on the camera's capture rate, which is set by the operator. Establishing the correct cadence of the footage relative to the final mastering frame rate, which can be 24, 25, 30, or 60 frames per second, is done in postproduction. In the field, all the frames rate at or below 24 fps, and at 25, 30, or 60 fps, will play back correctly from the camera's recorder. Frame rates above 24 fps, with the exception of those rates noted above, will not play back correctly in the field without using Panasonic's Frame Rate Converter or a desktop editing system capable of extracting and outputting the unique frames in the proper cadence for the operator's selected mastering frame rate.

Figure 5-9

Panasonic AJ-HDC27 VariCam integrated with filmmaking accessories.

DVCPRO-HD has a higher color sampling rate (3:1.5:1.5) and less compression (6.7:1) than HDCAM. The data rate recorded on tape is 100 Mbps, which is lower than the HDCAM format because this format has less resolution (a smaller frame size) than a 1080 progressive image. The 720p format balances its lower resolution by retaining more color information and requiring less compression to record on tape.

Sony's CineAlta camcorder has a prism-based imager with three 2/3-inch FIT 2.2 million pixel CCD sensors and 12-bit analog-to-digital conversion to capture 1920 by 1080 progressive images in the 16:9 aspect ratio. These images are subsampled to 1440 by 1080 for recording on HDCAM tape using Sony's PsF recording scheme. The CineAlta can also capture images at the following progressive frame rates: 23.976 fps, 24 fps, 25 fps, 29.976 fps, or 30 fps; and in interlaced mode at 50 fields per second and 59.94 fields per second. HDCAM is an 8-bit format, with 3:1:1 color sampling, a 7:1 compression ratio, and a data rate of approximately 140 Mbps.

Sony makes other HD cameras for field and studio use by broadcasters. Sony's HDC900 studio camera and HDC950 portable camera rely on the same imager as the CineAlta camcorder. The major differences are that the HDC900/950 can output an HD signal in either the 1080i, 1080psf, or 720p format and can downsample to output an NTSC or PAL signal. Sony's HDC930 is a less expensive interlace-only version, which will output a 1080/60i, 1080/59.94i, or 1080/50i signal.

Thomson's LDK 6000 mk II WorldCam has the identical prism-based imager and 2/3-inch CCD sensors used in the Viper camera. Three 9.2 million-pixel (1920 by 4320) HD-DPM+ (Frame Transfer) CCDs, 12-bit A-to-D conversion, and 22-bit digital signal processing permit capture of full-resolution HD—1920 by 1080 or 1280 by 720—images in either the 1.78:1 or 2.35:1 aspect ratio using Thomson's dynamic pixel management. The WorldCam can output the following formats to HDCAM, DVCPRO-HD, or D5-HD recorders: 1080psF at 23.98, 24, 25, or 29.97 fps; 1080p at 23.976 fps; 1080i at 50 or 59.94 fps; and 720p at 23.98, 25, 29.97, 50, or 59.94 fps.

The HDV Format

JVC and Sony introduced low-cost HDV camcorders during 2004 that record 1080i or 720p on mini-DV tape. HDV uses *interframe compres-*

sion—compression across multiple frames rather than within a single frame—to record 1080/50i, 1080/60i, or 720/30p at or below the 25 Mbps data rate of the DV format. HDV stores less than one-seventh the information on HDCAM and less than one-fifth of the information on DVCPRO-HD. To accomplish this feat, HDV relies on MPEG-2 compression.

MPEG-2 compression relies on both inter- and intra-frame compression to achieve greater efficiency than Motion-JPEG (M-JPEG), which is the approach taken by most digital videotape formats because editing is simpler when compression is applied to individual frames (intra-frame) rather than across multiple frames (inter-frame). MPEG-2 minimizes the bandwidth requirements by creating what are called long groups of pictures (GOP) and applying very high rates of compression. HDV use a 15-picture GOP that begins with an intra-frame (I-frame). The I-frame contains all the information to reconstruct a complete picture. It's similar to a frame compressed using M-JPEG.

The other 14 pictures are either bi-directional predictive frames (B-frames) or predictive frames (P-frames). B and P frames contain only information about the differences between the frames that come before and the frames that come after in the sequence. These 14 images only store partial information and consequently, require less bandwidth than I-frames. To see one of the B- or P-frames as an individual frame, however, the entire sequence must be decompressed to reconstruct that frame using the data contained in the I-frame and in the B- or P-frame.

For example, if you shoot an interview with the person sitting in front of a plain background, the background is not going to change very much from frame to frame. An HDV recording will have an I-frame that captures your interview subject against the background and then the next 14 frames will only capture his or her changes in facial expression and body position. The information about the background has to be recreated when you decide to use one of the B- or P-frames as an edit point. The difficulty of editing long GOP MPEG-2 compressed footage wasn't much of a consideration because MPEG-2 was developed for broadcast transmission, and not for production. For more on digital compression, see Chapter 4.

Professional HDV camcorder designs, using either prism-based three-CCD sensors or Bayer-filtered CMOS sensors, are just arriving on the market so it's too soon to know whether HDV will succeed as the low-cost alternative for HD production. **The Sony HVR-Z1U** is the first professional three-CCD camcorder to be released. JVC's professional HDV camcorder—a prototype displayed at the 2004 NAB conference—will like-

Figure 5-10

JVC Professional Products Company's ProHDV camera, on display at NAB 2004. (*Photo by Mark Forman.*)

ly use a CMOS sensor to capture HDV at 720p and is expected to be released soon (Figure 5-10).

Sony's HVR-Z1U uses a prism-based imager with three 1/3-inch SuperHAD CCDs in the 16:9 aspect ratio. The camcorder records in HDV or DVCAM at 60i or 50i and also at 30, 25, or 24 frames per second, although it's not in true progressive mode (Figure 5-11). Sony approximates the look of progressive using what they call CineFrame; no details

Figure 5-11

Sony's HDV camcorder is a milestone in low-cost HD acquisition. (*Photo by Mark Forman.*)

are available as this book goes to press regarding how this function works. Two CinemaTone gamma-correction curves intended to approximate film's gamma response to light, are selectable. The HVR-Z1U has a color correction feature that provides controls to adjust the color phase and saturation of up to two specific colors. Among the other interesting features are six user-assignable function buttons, and a viewfinder that is switchable between black and white and color.

Progressive Standard-Definition Cameras

Panasonic pioneered progressive scanned SD camcorders. Three Panasonic camcorders are currently in use for digital cinema purposes.

Panasonic's AJ-SDX900 uses a prism-based imager—16:9 and 4:3 switchable—with three 2/3-inch 520,000 progressive-capture, native 16:9 IT CCDs and 12-bit digital signal processing. The camera section captures images at 60 fields per second or 30 or 24 frames per second. The SDX900's recorder can operate at 25 Mbps in the DVCPRO format or at 50 Mbps in the DVCPRO-50 format.

DVCPRO is Panasonic's professional version of the DV format and uses 4:1:1 color sampling. DVCPRO-50 doubles the data rate and uses 4:2:2 color sampling. Both of these formats are recorded at 60 fields per second. When the camera is operating in a progressive capture mode, the progressive image is recorded into both the odd and even fields that are normally used to make an interlaced frame. Operators have two choices of how to record 24-frame progressive images on tape. The industry standard pull-down arrangement replicates the approach used to transfer 24 frame per second film to 30 frame (60 field) per second NTSC video. Panasonic's advanced pull-down arrangement was developed to make extracting 24 frames from a 60i recording easier using desktop video editing software.

Panasonic's AJ-SPX800 records on solid-state memory called P2 (in the shape of a PC-MCIA card) instead of tape, using the same prism-based imager as the SDX900, although with 14-bit instead of 12-bit digital signal processing. This 16:9/4:3 switchable camcorder has three 16:9, progressive, 2/3-inch 520,000 pixel IT CCDs. Operators can select to record at 25 Mbps with 4:1:1 color sampling in the DVCPRO format or at 50 Mbps with 4:2:2 color sampling in the DVCPRO-50 format. Up to five P2 memory cards, which are hot-swappable, can be inserted into the camcorder. The camera can capture at 60 fields per second or 30 or 24 frames per second. Only the actual frames captured by the camera are stored on the P2 memory cards. Consequently, only one pull-down arrangement—the standard one—is needed to output 24-frame material.

Panasonic's AG-DVX100A uses a prism-based imager with three progressive 1/3-inch 410,000 pixel IT CCDs and has a fixed lens (Figure 5-12). This prosumer camcorder captures at 60 fields per second or 30 or 24 frames per second and records in the mini-DV format at 25 Mbps. Operators have two choices of how to record the 24-frame progressive images. The industry standard pull-down arrangement replicates the approach used to transfer film to NTSC video. Panasonic's advanced pull-down arrangement was developed to make extracting 24 frames from a 60i recording easier using desktop video editing software.

Canon's XL2 prosumer camcorder uses a prism-based imager with three progressive, native 16:9 460,000-pixel CCDs and accepts interchangeable lenses with an XL mount. The CCDs have a diagonal of .289 inch (smaller than 1/3 inch) in the 16:9 mode and a .236 inch diagonal (smaller than 1/4 inch) in 4:3 mode. The XL2 captures images at 60 fields per second or 30 or 24 frames per second and records them in the mini-DV format at 25 Mbps. Operators have two choices of how to record the 24-frame progressive images: the industry standard pull-down arrangement or the advanced pull down-arrangement developed by Panasonic. This camera is notable because of its innovative design and use of interchangeable lens. Adapters are available to mount Canon still camera lenses or Nikon still lenses on the XL2.

Figure 5-12

Panasonic's AG-DVX100 24p mini-DV camcorder.

Sony's MSW900 camcorder has a prism-based imager with three 2/3-inch Power HAD EX one million pixel CCD 16:9/4:3 switchable sensors, and 12-bit analog to digital conversion. This camcorder records images in the MPEG IMX format. This is an MPEG-2 compression format that has I-frames only, 4:2:2 color sampling, and a data rate of 50 Mbps. The NTSC version of this camcorder is capable of recording 60 fields per second or 30 progressive frames per second. The MSW900P (the PAL version) can record 50 fields or 25 progressive frames per second.

Sony's PDW-530 uses the identical imager as the MSW-900, but records on optical disks in either the MPEG IMX or DVCAM formats. DVCAM is Sony's professional version of the DV format and has a data rate of 25 Mbps. There are three selectable levels of compression available when recording in MPEG IMX format. Users have the choice of a data rate of 30, 40, or 50 Mbps. With the CBK-FC01 option installed in the PDW-530, this camcorder can capture 24 progressive frames per second in addition to its normal frame rates of 60 fields or 30 progressive frames per second, for output to its optical disc recorder.

Interlaced Standard-Definition Cameras

There are dozens of models of interlaced cameras and/or camcorders, for broadcast, cable television production, news gathering, or industrial recording purposes that can be used for digital filmmaking. The images output by these cameras need to be wde-interlaced in postproduction. The advantage of these cameras, however, is that they are plentiful and—consequently—often the least expensive option open to an indy filmmaker.

If you have a choice, PAL cameras are the better option. PAL records at a frame rate of 25 fps. This means that after de-interlacing the images, each frame can be transferred to a 24-frame progressive format or to film, one for one. The finished show will run four percent longer than it does in PAL. The sync sound is slowed four percent and pitch-shifted to bring it back in sync for projection at 24 frames per second. The smarter choice is to avoid cameras that can only record an interlaced signal and select one of the ones that does progressive scanning.

And whatever you choose, it will be your own creative talent that will make the real difference in whether what you shoot is or isn't great cinema.

Postproduction

Film or digital, once a motion picture has been photographed "postproduction" begins. This umbrella term refers not only to the editing of the film—the splicing together of separate shots to tell a story—but also to all the other things that are done to the movie in the last stages of its creation prior to it being distributed and exhibited in theaters. Postproduction includes not only editing the images, but creating sound effects and re-recording dialogue, editing the soundtrack—including the music—adding visual effects, and finally color-grading the end result from which release prints will be made.

Because the postproduction phase of a motion picture (or of a television show, for that matter) is so complex and multi-faceted, this stage ordinarily takes much longer than the time required to actually shoot the images. A film may be photographed in two months; editing can take twice that long. The popular image of making a movie is one in which the director is on the set, the cinematographer is behind the camera, and the actors are in their costumes. The director, the cinematographer, and the actors perform their respective tasks and a movie is made. In reality, however, simply photographing a movie is only the beginning of the filmmaking process. Next comes postproduction, the crucial stage during which all the "raw materials" that the director, cinematographer, and the actors have created are combined with other content—sounds, music, visual effects—and crafted into the film that eventually is shown in theaters. Postproduction also requires the services of a cadre of talented artists; first and foremost is the editor, but there's also the sound editor, the composer, the Foley artist (for sound effects), visual effects supervisor and visual effects artists, the color timer, and other talents.

The fundamental steps of postproduction remained unchanged for decades. Had Thomas Edison or the cinema's other early pioneers been able to witness a movie being edited in 1980, much of what they would have seen would have still been recognizable to them. Ten years later, that would not have been the case. What made the difference was the arrival of digital cinema technology. New digital postproduction technologies have significantly impacted moviemaking, not only introducing new systems, but entirely new workflow processes. This is true of $100 million Hollywood blockbusters and also of low-budget independent movies. Why has this happened? Because these new digital cinema postproduction technologies improve the tools of creative expression, provide better control over the moviemaking process, and—in some cases—even save money.

A Brief History of Film Postproduction

In the early days of filmmaking, movies resembled short stage plays. The earliest films were quite short and only involved a handful of scenes at most. The uninterrupted action enabled films to be rapidly produced to meet demand. At some point, the director—who had to keep to a tight deadline—realized that if there was a mishap on stage, he could simply stop the camera and start the action again when the problem was sorted out. Once developed and projected, the film would reveal no break in continuity.

Directors soon realized, however, that stopping and repositioning the camera could make the story more interesting, and it became an accepted practice. But editing was still mainly a matter of trimming ends of film rolls and splicing "takes" together. In 1903, director Edwin S. Porter's 10-minute long, 14-scene *The Great Train Robbery* introduced a number of innovations, including the first use of title cards, storyboarding the script, a panning shot, and cross-cutting editing techniques. Jump-cuts or cross-cuts were new to viewers. With this technique, Porter showed two separate lines of action or events—the actions of the bad guys and the good guys—happening continuously at identical times, but in different places. Other films credited with further developing the sophistication of editing are D.W. Griffith's controversial *Birth of a Nation* (1915) and Serge Eisenstein's *The Battleship Potemkin* (1925). Editing evolved from the labor of cutting and splicing film footage to an articulate language of storytelling.

Along with the development of editing as a craft and art came new tools to make more sophisticated editing possible and help speed up the process. Around 1917, Dutch immigrant Iwan Serrurier got the idea that a home movie projector enclosed in a wooden cabinet, like a Victrola, would be welcomed by studio executives who could then view "dailies" (the raw footage shot each day during a film's production) in the comfort of their offices. In 1923, Serrurier manufactured about 15 of these machines, which he dubbed the Moviola. Though he only sold three machines from 1923 to 1924, Serrurier had a fateful meeting with an editor who thought the Moviola could be adapted for use in the edit room. At that time, film was examined over a light well, spliced, and then run in the projection room; it was an iterative and very time-consuming process, especially since the editor had to imagine the images in motion. Serrurier removed the projection lens and lamp house, turned the machine upside down, and attached a viewing lens. Editors loved the hand-cranked Moviola, which allowed them to see frames in motion.

Soon Hollywood editors uniformly adopted use of the Moviola, from Warner Bros. to MGM. Serrurier later built the Moviola Midget, powered by a sewing machine motor. With the advent of sound, Serrurier built sound heads for optical sound, as well as viewers for 16mm and 35mm and the early 65mm and 70mm films. The Moviola worked with an intermittent shutter mechanism, but had some drawbacks. The image that the editor saw was small, the shutter mechanism was noisy, and—since the editor fed in the film by hand—it couldn't go faster than double speed.

The next evolution in film editing was the development of so-called "flatbed" film-editing machines, which enabled the editor to sit down while he or she edited, accommodated various combinations of sound and picture heads, and was less noisy than the Moviola. Kems and Steenbecks

Figure 6-1

Apple Computer's Final Cut Pro sets a new standard for powerful, yet affordable nonlinear editing on the Macintosh computer.

allowed the editor to look at the film on a bigger screen in 10-minute chunks, thanks to the use of a rotating prism, which made operation relatively quiet and enabled playback in faster-than-normal speed. The Steenbeck and KEM flatbed editing devices were introduced and soon became heavily used. The use of these flatbed editing systems continued to be the norm in postproduction until the introduction of digital editing systems such as the Avid.

A Brief History of Visual Effects

Visual effects have been part of motion picture history almost since the beginning. In 1895, *The Execution of Mary, Queen of Scots* created the illusion of Mary losing her head on the chopping block. The director stopped the camera in mid-scene, the actress was replaced by a dummy in identical costume, and the camera was restarted as the dummy had its head chopped off. Viewed as a continuous sequence, it looked convincing for its time. It was, however, the brilliant French magician-turned-cinematographer Georges Méliès who was the first to ambitiously exploit "in-camera" effects in more than 500 films made between 1896 and 1912. Staged with the immobile cameras of that era's filmmaking, Méliès combined a magician's mechanical tricks with photographic techniques to create magic on celluloid. His "trick" cinematography included masking, multiple exposures in camera, miniatures, and other techniques—some of which are still used to this day.

Méliès was soon joined by other early cinematographers who enjoyed exploring "trick" photography to portray impossible imagery. Robert W. Paul developed a printer that allowed him to combine images from several negatives. Edwin S. Porter, director of *The Great Train Robbery*, also made trick films with miniatures and stop-motion (or "arret") as well as mattes that combined shots of a moving countryside behind a stationary train car to make it appear that the train was moving.

Filmmakers became increasingly inventive with camera techniques and miniatures to create imagery that wowed the audiences. Matte paintings, puppets, miniatures, and models all played a role in these early decades of filmmaking.

With the advent of sound, effects became even more realistic. Now there was an aural complement reinforcing the visual illusion (imagine Mary Queen of Scots being beheaded with the sound of a loud "Chop!" at the crucial moment). Rear projection was developed in the early 1930s to add authenticity to backgrounds. Perhaps most important of all was the development of the optical printer into a working device. Though the basics of an optical printer—which married two or more moving images into one—was first developed in film's early days, it took "talkies" and improved film stocks to bring it into serious play in Hollywood. Within a decade, optical printers had grown in sophistication, with all the accessories required to create complex and realistic imagery. The viability of the optical printer sparked a boom in the use of scale models and miniatures. Footage of an actor in a dinosaur suit stampeding a miniature city could be combined in the same frame above film of extras running toward the camera for a relatively convincing depiction of citizens fleeing a monster. Within short order, filmmakers were able to avail themselves of a range of sophisticated techniques for creating visual effects, either in camera or, as a postproduction process, with the optical printer.

Video Changes the Landscape

As noted in Chapter 1, the Ampex Corp. created the first commercially practical videotape recorder (VTR) in 1956, which captured live images from television cameras by converting visual information into electrical impulses that were then stored on magnetic tape. CBS and then NBC began to use Ampex's VR-1000 VTR, with the rest of the industry quickly falling in line. By 1962, Ampex also introduced its first system to edit video, but it wasn't until CBS and Memorex created a partnership that video editing became a widespread reality. Their partnership, dubbed CMX, resulted in the CMX 600 editing system and the CMX 200 online editing system. For the first time, editing was based on timecode, by which each frame was identified by a number representing its position in time. Also born was the idea of doing "offline" or creative editing on a less expensive editing system and then "conforming" the results with an "online" editing system.

Television led to the development of two different postproduction tracks. Whereas feature films continued to follow the time-tested track of physically cutting film and gluing shots together, television editing slowly but surely moved to video editing and video postproduction. The development of the telecine enabled producers to transfer moving images from film footage to videotape and then edit those images electronically as video. The development of different videotape formats—including Sony's workhorse 3/4-inch U-Matic tape format and its Betacam professional half-inch tape format—also ensured the acceptance of videotape, not just as a post-production medium, but also for acquisition. In the field, the Betacam camcorder began to replace the 16mm camera for news acquisition by the late 1970s. At the time, this had little impact on theatrical filmmaking—but it would plant a seed for today's digital postproduction revolution.

Videotape was an excellent medium for time-delayed playback and other requirements of television stations and networks. But there were significant trade-offs. Videotape turned editing into a linear process. Film, as a physical medium, is inherently "nonlinear," which means that the editor can cut any number of frames from one part of the roll and insert it anywhere else in the film—over and over again, without penalty. Videotape changed that dynamic. As electrical pulses recorded on magnetic tape, the editor could no longer cut and splice. If an editor (or, more likely, a producer) decided to change something 58 minutes into a 60-minute program, the videotape process required that the entire program be re-recorded or re-created in order to get to the point where the change needed to be made.

The early days of videotape also introduced video devices that enabled basic effects and generation of type (these were called "character generators"). The different videotape formats improved resolution, representation of color, and ease of use. By the late 1980s, Sony had introduced D-1, an uncompressed component digital videotape format with better picture quality than any video format that had gone before, and Ampex had debuted D-2, a composite digital video format that was a close second and cost less. Digital video seemed the perfect answer for those who decried the fact that, with every video edit, a new "generation" of the picture resulted and video picture quality suffered. As noted in Chapter 2, digital video is just zeroes and ones, so quality is theoretically unchanged.

Then Along Came Digital

In the mid 1980s, editing manufacturers had already begun to experiment with creating new editing systems that could return the advantages of "nonlinear" editing to video. Tape-based, optical-based, and hybrid systems appeared in the market as pioneering systems from Montage, Ediflex, CMX, and—from George Lucas' R&D lab—EditDroid.

The idea was bold, but technology wasn't yet available to support the idea of electronic nonlinear editing. The quality of the picture and the difficulty and expense of digital video storage were two big obstacles. To handle these issues, the early manufacturers focused on different compression schemes—ways to reduce the amount of data that represented each image—that would enable the best quality picture image with the least amount of storage.

Numerous manufacturers of nonlinear editing systems came and went, with companies often being acquired by one another and having all or part of their technology incorporated into the surviving brand's newest model. Most notable of these early companies and a leader in the field of nonlinear editing to this day is Avid Technology, which introduced its first nonlinear editing system in 1988, on an Apple Computer Macintosh CX. The first unit shipped toward the end of 1989, and by 1991, Avid became the first manufacturer to offer JPEG compression running full 30 fps video.

Editing Machine Corp. (EMC) was another pioneering manufacturer of nonlinear editing systems. Founded in 1988, EMC shipped its first nonlinear digital system in 1988, the first to use Motion JPEG as its compression technology, and the first to use magneto-optical discs for storage. Lightworks was founded in 1990 in the United Kingdom and offered a system designed to be easy to use for editors accustomed to film editing interfaces. Based on the Steenbeck's jog/shuttle knob control, Lightworks began shipping in the United States in 1992. Axial and Accom, meanwhile, merged in 1991 and launched RAVE, a hybrid linear/nonlinear system offering 10-bit recording in D-1 and D-2 with no compression. ImMIX debuted its VideoCube digital video workstation in 1993, the same year that Data Translation launched its Media 100, which operated on the Macintosh.

From the beginning, these digital nonlinear editing systems offered more than just editing. Their inventors realized that the digital environment made it possible for several tools to be offered as software within the same box. Character generation, video "paint" graphics, and audio mixing made their way into some of the very first nonlinear editing systems.

But the film and TV industry balked at using nonlinear editing systems. The cost of storage still made them expensive, the software was still buggy, and the computer platforms were slow and unreliable. At the crux of their reluctance were two major factors. Many real-world users were reluctant to give up tried-and-true technologies when it came to meeting deadlines and producing product. And, in the early 1990s, most filmmakers and editors had limited experience with working with computers of any kind.

It took some brave pioneers to give the new nonlinear editing technology a try. During the 1990s, a few high-profile filmmakers, such as George

Lucas, took the leap and their exploits with digital nonlinear editing served as an example to everyone else waiting on the sidelines. At the same time that a handful of filmmakers were taking the challenge of being first to use these less-than-perfect systems, technology continued to improve. As noted in Chapter 2, however, steady improvements in the price/performance of computer processing technology—as noted in what is called Moore's Law—gradually improved the nonlinear editing proposition.

Within a decade, the cost and capabilities of the computer platform, digital storage, and software sophistication had caught up with the needs of the industry. The use of digital nonlinear editing systems—even as the number of manufacturers declined—became widespread, both in the editing of feature films and television programs.

Editing wasn't the only task to go digital in the 1990s. Sound editing and mixing also benefited dramatically by the move to digital toolsets. Visual effects also evolved from analog to digital. The inherent "trickery" of a visual effect requires sophistication to fool the eye, which meant that visual effects artists needed a higher level computer processing than a desktop PC, and they needed to write their own software to enable very specific tasks. A handful of software manufacturers—Alias, Wavefront, TDI—came to the fore as did computer workstation manufacturer Silicon Graphics (SGI).

Visual effects for feature films had other daunting requirements. To add a digital element to a film frame meant that the film information had to somehow be brought into the digital realm. Once the film information was married with digital elements, the end result had to be recorded back out to film. In the early 1990s, there were no commercially available devices to do either of these tasks. Early visual effects pioneers simply had to create these devices themselves. And they did. The necessity to create tools meant that very few companies provided digital visual effects. Early, innovative work in computer graphics came from Robert Abel & Associates, III, MAGI, Digital Effects, and Omnibus Computer Graphics. To make *Star Wars*, George Lucas and a handful of visual effects pioneers established their own studio and their own processes. Although visual effects companies of those early days are largely gone, the alumni of those companies are still leading figures in today's visual effects community.

The technology for transferring digital images to film has advanced greatly as well, with companies such as CELCO and ARRI today providing extremely sophisticated systems for this purpose (Figure 6-2).

Figure 6-2
Leading motion-picture camera manufacturer ARRI's latest technology to address the digital intermediate market is ARRISCAN, a 16mm and 35mm film scanner that turns film images into digital data with a resolution of 6,000 lines ("6K") per frame.

Democratization of Digital

The continuing impact of Moore's Law made computer platforms faster and less expensive—to the point where ordinary personal computers became more than adequate as engines for editing and visual effects work. As the market grew, innovative software authors—sometimes university students or aficionados working out of a garage—wrote tools to enable editing, compositing, audio mixing, animation, and graphics on Macs and PCs. As the base of the pyramid expanded, expensive proprietary computer systems began to be replaced by ordinary desktop computers, even at the most established visual effects companies. As digital technology became a commodity—and not a proprietary and exclusive process—the number of editors and visual effects artists using digital tools exploded. By the mid 1990s most television shows were edited with nonlinear editing systems—usually the Avid Media Composer—with numerous motion-picture editors also using the Lightworks editing system.

Although storage was still expensive and the adoption of digital technologies posed infrastructure challenges, the evolution of computer technology meant that increasing numbers of talented people could try their hand at offering digital editing, sound mixing, or visual effects. As a result, the boutique studio—or "professional project room"—was born. Individuals were able to open postproduction businesses in their homes, thanks to low-cost digital technology.

At the same time that less expensive hardware and software made it possible for more people to offer digital postproduction services—often at competitive prices—the distribution of entertainment was also being impacted by new digital technologies. In the United States, the broadcasting industry, computer makers, and the government devised a plan for transitioning the nation to digital television (DTV), a new broadcasting format to replace the analog NTSC system in use since the mid 1950s. To accommodate multiple interests, DTV includes use of multiple formats, including a digital version of high-definition television (HDTV).

HDTV doubles the number of lines (the "resolution") that make up a television image for a much clearer, more detailed image, and also provides improved (digital) sound and a wide cinema-style screen, among other benefits. The U.S. Congress has mandated that broadcasters will switch from NTSC to DTV at some point (the current target date is 2009), although the actual transition may take far longer to accomplish due to the cost of replacing existing infrastructure and the huge installed base of standard-definition (SD) consumer NTSC television sets.

Nevertheless, the eventual necessity to produce television programming in HD has upped the ante for postproduction houses offering a range of analog and digital services. For those facilities serving the TV market, the need to be able to edit, mix sound, and offer character generation, graphics, and effects in the HD standard became an imperative. Though the vast majority of consumers have analog TV sets (which continue to be sold), the TV programmers and their postproduction partners are already hard at work making a switch to HD so that the programs they make today will be still be salable in the future.

Film Is Digital

Despite numerous high-profile examples of Hollywood directors photographing their movies with digital HD tape (see Chapters 3, 10, and 14) and the large number of independent filmmakers using digital HD or SD camcorders to shoot their pictures (see Chapters 11, 12, and 13), mainstream theatrical movie making is still dominated by 35mm film. Benefiting from more than a century of continually advancing chemistry, today's film stocks offer greater resolution than current digital HD imaging, although the performance gap between the two media is narrowing. Ironically enough, however, advances in digital postproduction technology are enhancing film's role as an acquisition medium, as opposed to threatening it. The reality is that motion-picture film is as much a part of the digital cinema revolution as digital HD or SD video.

Leading-edge film-scanning technologies—such as the IMAGICA Imager XE, the ARRISCAN, and the Thomson Spirit 4K DataCine—provide a very effective "gateway" for film-acquired images to be turned into digital data. These data can also be output back to 35mm film (for creation of release prints for theaters) via the CELCO or ARRI Laser systems (Figure 6-3). In between, however, is a realm in which film-acquired, digital moving images can be modified and manipulated in an infinite number of ways. This realm is known as the *digital intermediate* (DI), a major part of which involves color grading—or "timing"—the image.

Traditionally, in the last part of the filmmaking process, the cinematographer works closely with a photochemical timer, who achieves the color

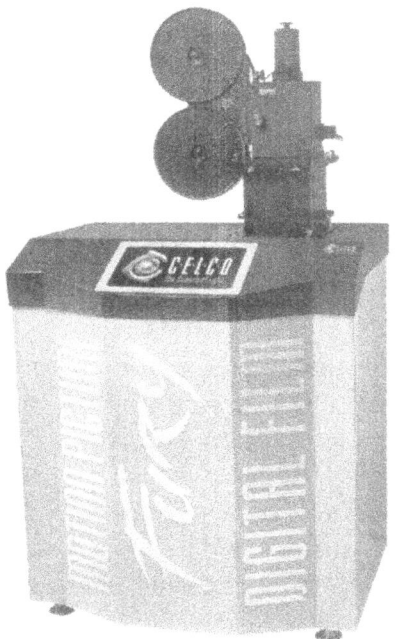

Figure 6-3

CELCO's FURY system is a digital film recorder that prints—frame-by-frame—digital image data of up to 4K resolution to any type of film stock, including 70mm. When the digital intermediate process is complete, CELCO's FURY is a leading choice for transferring film-originated and/or digitally generated images back to film again.

and contrast desired by tweaking the chemicals in the "bath" in which the film negative is developed. But after digital visual effects demonstrated the power of enhancing and changing an image in the digital realm, digital color tweaking became the next logical step. In just a few short years, DI has become very popular among filmmakers, who sacrifice nothing in terms of the integrity and beauty of their original 35mm-acquired images. If anything, digital color timing offers more creative control and color precision than traditional photochemical technologies ever did. And when the DI process is completed, CELCO and ARRI Laser systems can deliver a pristine 35mm copy of the finished product.

The declining cost and increasing quality of film scanning technology has encouraged a growing number of directors and cinematographers to embrace DI. And once in the digital domain, the increased creative latitude that's possible has even encouraged some filmmakers to make radical and specific changes to their images that were impossible with photochemical color timing. In the feature film *O Brother, Where Art Thou?* (2000), for example, cinematographer Roger Deakins, ASC, was able to change the color of certain scenes using the DI process to create a sepia-tone look that enhanced the mood of the film's Depression-era story. Though currently only a healthy minority of films go through the DI process, most industry observers believe that nearly all films will receive this treatment within the next few years.

The very distribution of film is also seeing a digital twist on the horizon. An alternative, called the Universal Digital Master, is a compelling concept in a business in which "films" are assets to be distributed in a variety of ways (in theaters, home video, pay-TV, and cable and broadcast networks). A Universal Digital Master is a digital tape that can be readily copied for distribution to and within motion-picture cineplexes, and also used for all ancillary media, from DVD to airplane release. It is also a high-quality replacement for reels of more costly 35mm film, which wears out after several months and must be destroyed to prevent piracy.

Digital motion-picture distribution and exhibition is still in its early days, as different companies compete with digital cinema systems, the studios express concern over security, and the directors and cinematographers worry about artistic integrity. The Hollywood film industry is basically a conservative environment that adapts to change very slowly. Digital cinema has the potential to disrupt many time-honored Hollywood business practices, and—as such—is unlikely to be adopted overnight. Nevertheless, nearly everyone agrees it's the wave of the future. How digital motion-picture distribution and exhibition might impact the postproduction processes remains to be seen. The case study of Hollywood DI facility EFILM, which follows this chapter, may well provide the best available answer to this question.

As For Now...

Feature filmmaking stands at a crossroads, as old techniques and processes are disrupted and the future still comes into focus. In many ways, postproduction is the most stable and established point in the digital process. Postproduction houses and postproduction artists have very successfully adapted to digital technology for more than 15 years now. Editors watched their jobs go from razor blades and splicing to keyboards and computer screens. Visual effects artists abandoned their optical printers for compositing software. Postproduction has experienced weathering the change to digital processes and equipment.

The next step will doubtless be a continued and increased evolution of a postproduction industry that's already gone digital. The change that is currently in the making is software programs—such as Apple's Final Cut Pro—that allow one person to accomplish a variety of tasks, from editing, to sound mixing, to visual effects. Already, the editor is able to rough out graphics and sound mixes. The visual effects artist goes on location to advise the cinematographer and director on how to best shoot what is known as the "background plate," the scene against which computer-generated imagery will be composited. Postproduction equipment offers the capability of an increasing number of tasks, and each postproduction artist is required to learn jobs that were once outside his or her purview. At the same time, the lines between what is "production" and what is "postproduction" become blurred, especially for visual effects. When a film includes an entirely digital character created after principal photography—such as Gollum in *The Lord of the Rings* trilogy—is it production or postproduction?

Another current trend is the networked postproduction facility and visual effects house. With all the computers linked by a network in a facility, multiple artists can work on the same footage at the same time. The network extends to other post-facilities and effects houses that may be working on other portions of the same feature film. The network now connects the director, the cinematographer, and a wide range of artists in Hollywood, across the country, or even across the globe.

The growing importance of digital technology is having the unintended consequence of bringing the postproduction artists into the limelight. The contribution of editors, sound mixers, and visual effects artists to the story-telling process has become more pronounced than ever before and certainly more evident to those on the production side of the equation. Many postproduction experts would argue that it's high time this happened.

EFILM: A Case Study in Digital Intermediates

Digital technologies have been improving the motion-picture postproduction process for more than a decade. Nonlinear editing, computer graphics and compositing, digital color correction, and high-definition film scanning and recording have all improved the creative control filmmakers have over the final look of their 35mm-originated images. As computer storage continues to grow in capacity and decline in price, the ability to "warehouse" all the uncompressed data of a theatrical motion picture while it's being creatively modified and manipulated becomes increasingly feasible. This process, generally referred to as "digital intermediate"—in which original film-acquired images are scanned into data and then artistically processed in what is literally an intermediate step before being output to film again—has become a powerful new creative tool.

EFILM, a Hollywood-based facility, provides a good "snapshot" of the current state of the digital intermediate (DI) process, as of 2005. According to EFILM's Vice President of Corporate Development Bob Eicholz, "What EFILM does is acquire images, make them look better, and then deliver them in one format or another. That's our business."

The simplicity of Eicholz's statement belies the leading-edge technology and highly sophisticated workflows that EFILM has evolved during its nine-year history (Figure 7-1). It's a combination that's empowering filmmakers with an unprecedented array of new visual storytelling tools while establishing the company as one of the world's leading practitioners of DI, a term Eicholz defines by the sum of its parts.

Figure 7-1

EFILM President Joe Matza (left) and VP of Corporate Development Bob Eicholz in one of the Hollywood-based company's four digital color-timing suites. (*Photo by William Norton Photography.*)

"The digital intermediate process is really all of our core-business service lines wrapped into one, and it involves entire films instead of just small pieces of them," he explains. "One part of that core business is film scanning, usually for production companies, postproduction facilities, and visual effects houses."

Full Capabilities

EFILM owns three IMAGICA IMAGER XE film scanners, which convert 35mm film frames into 10-bit RGB image data at the rate of one frame every four seconds for 2,000-line (2K resolution) imaging and double that for 4K. On the other end of the DI equation, EFILM is equipped with 13 ARRI Laser recorders to print finished digital motion-picture data back to 35mm film.

"We have the world's largest installation of ARRI Lasers," Eicholz continues. "Scanning clients return to us once they've finished creating their effects so we can record their material back to film. That film then goes to the lab for processing of negatives ready to print."

"Tape to film is another flavor of the same thing," he adds. "Typically someone will originate material in some type of video format—not film—and it arrives here on a tape. We'll do some type of image processing to make it compatible with film formats; we can also improve the color, reduce the noise—lots of things to make it look better. And then it goes on to our ARRI Laser recorders and the lab for making negatives that can then be used to print and distribute. A lot of these are small Sundance filmmakers."

"There are also large films—such as Robert Rodriguez's *SpyKids 2* and *SpyKids 3D*—which originated on 24p HDCAM tape," explains EFILM President Joe Matza. "We made seven film negatives for that particular project, which allowed Rodriguez to release first-generation prints and avoid the traditional laboratory-duping process."

EFILM has a decided advantage in this area; after five years of private ownership, camera-rentals giant Panavision acquired the company in July 2001. One year later, Deluxe Laboratories bought 20 percent of EFILM. In 2004, Deluxe took over full ownership of the company.

"The neat thing about this is that we now have within our 'family' the whole 'life cycle,' starting with initiating and filming and on to DI, all the way through to lab services and worldwide distribution," Eicholz enumerates. And although EFILM's clients may also happen to work with other camera suppliers and labs, the company's "family" relationships give clients a definite edge.

"The really important part of this is our link with the lab," Matza states. "It's something we've worked on for two years. If you think about the photochemical process, the fact is that the mix at the lab changes every day. So every day—and sometimes more than once a day—we send special film there to sample their chemicals, and then we tweak and tune our film recorders so that the recording that goes on that day matches the chemicals in the lab. Some of the companies that are just now trying to get into doing digital intermediates will be in for a rude awakening if they assume that just sending a digital file to the lab will mean that the color's going to be right."

The Science of Color

In between film scanning and film recording is color timing, where much of the magic and creative power of the DI process resides. Whereas motion picture "color timing" (or "grading") has traditionally been a laborious procedure ever at the mercy of the vagaries of photochemistry, digital color timing is far more efficient in terms of speed and exactitude. Today's digital color-timing systems provide a wealth of new creative abilities borrowed from the tools and techniques of the long-established world of standard-definition video color-correction. This includes the ability to isolate or "window" specific areas of an image and change its color or brightness while leaving surrounding areas unaffected (Figure 7-2). Working with directors and DPs, master EFILM colorists such as Steve Scott have made these revolutionary new tools integral to the unique looks of such stylish 2004 releases as *Van Helsing* and the remake of *The Lady Killers*, and the blue-screen-intensive *Sky Captain and the World of Tomorrow*.

EFILM has four color-timing suites (with additional rooms planned); each is equipped with a Silicon Graphics Onyx2 supercomputer-based color-timing system running proprietary software developed in collaboration with ColorFront, original developers of the Discreet lustre system.

Figure 7-2

EFILM owns three IMAGICA IMAGER XE film scanners, which convert 35mm film frames into 10-bit RGB image data at the rate of one frame every four seconds for 2,000-line (2K resolution) imaging.

"One of EFILM's principal competitive strengths is the work we do with our color scientist, Bill Feightner, and others to enhance and understand color," Matza offers. "He has developed proprietary look-up tables [LUTs, software that translates or maps the colors produced by one type of display, such as a digital projector, to those rendered by another, such as projected 35mm film]. We also do special modifications to our projectors and other software enhancements to ensure that what you see digitally on the screen is what you'll see on film when it comes back from the lab. Our goal here is that the digital file goes to the lab, gets printed, and it's done."

In addition to proprietary color-timing systems, each EFILM DI suite is equipped with both a 35mm film projector and a Texas Instruments DLP Cinema ("black chip")-based digital projector. EFILM is in the midst of converting all of its projectors from 1K to 2K models manufactured by either Barco or Christie.

"We like TI's DLP technology," Matza asserts. "It's the most reliable and produces the best blacks."

"When you see 2K projection, it looks great," Eicholz agrees. "The contrast looks great, the color's great; I don't think any moviegoer would complain about seeing an image from a 2K projector versus film. One of the other great things is the steadiness of the image in 2K. Even the best film projectors have some weave in them, but digital is rock-solid. There's something very captivating about an image that's not moving at all. It's more accessible. I think all of these things, over time, will change the way that cinematographers think about film" (Figure 7-3).

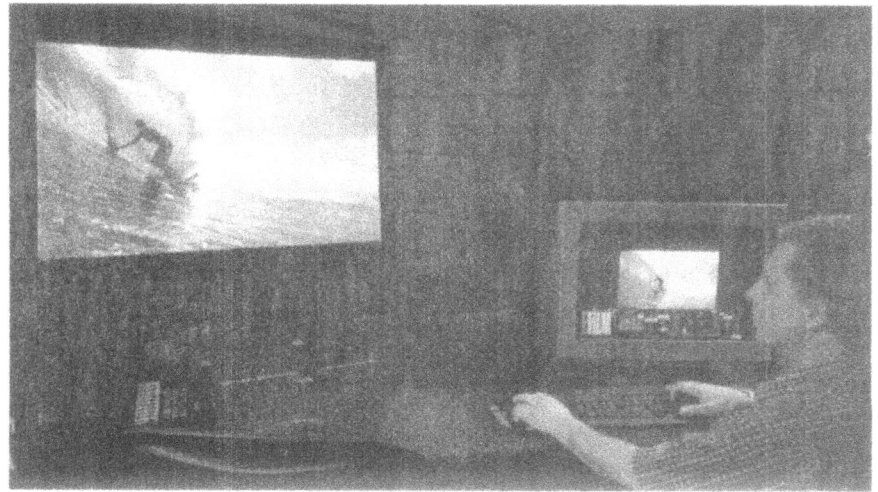

Figure 7-3

EFILM's proprietary color-grading system at work on the motion picture *Blue Crush*.

New Workflows

EFILM's DI capabilities not only are expanding the palette of "looks" directors and DPs can apply to their films, they're also changing the very workflow of the motion-picture postproduction process itself. Once a film's original footage is scanned and becomes a body of high-resolution digital data, new creative and workflow options arise.

"We do dust-busting," Eicholz elaborates. "We have a team of people and automated systems that identify dust particles and get rid of them. Digital intermediate also includes what we call the digital mastering process. We do titles, digital opticals—fades, dissolves—the conforming, the online assembly, so you really don't have to cut your negative."

"And these are high-resolution files—not data files—we're talking about," Matza interjects. "These are 2K, 4K, 3K resolution files and not just online assembly in the 'video' sense."

"When our customers first came to us, their films were completely cut negatives," Eicholz continues. "This is the way we did *We Were Soldiers*, which was the world's first 2K digital intermediate and the first that was 100 percent digitally color timed in an all-software platform. We scanned the whole thing in, color-corrected it, and put it on film. We scanned it at a true 2K; in fact, the film was double oversampled at 4K and then converted to 2K images, which is the best way to actually acquire film images and digitize them. This helps assure higher-quality 2K scans without causing the storage and I/O obstacles that come with working with features scanned at 4K."

"To create a 2K image, you've got to scan at 4K and downsample," Matza explains. "We do all of our movies that way. It's a good thing for the industry right now, to scan at 4K, downsample at 2K, do all your processing, and then take it back to 4K. There's a subjective gain on the output and there's a very practical gain on the input. There's also nothing wrong with doing *everything* at 4K except that it's four times the data, four times the resolution, and four times the time. We're working on the world's first 4K feature, which is in our shop right now. It's a massive amount of storage and an expensive proposition. But three years from now? A 4K digital intermediate probably won't be that costly."

Conforming

"Today we're getting original camera rolls and edit-decision lists; we have software that does an autoconform, just like you would in the video world with an Avid," Eicholz says. "We take an EDL and assemble an entire film. Sometimes we will reassemble a new edit of the film along the way because films can change up to the last minute. So even though we may have started color-timing, we may need to do some edit changes, which will lead to digital cinema previews.

"Related to that is something we've done on a recent film. Instead of just doing one conform—one edit—we did four. Then we output digital cinema versions that are taken and shown to audiences. Audience feedback is gathered, editors make a new EDL, we reconform it here, color-time it a bit more, and we can go through that process again if need be. This works very well, to be able to show a beautiful digital print in preview screenings. We see this as an unexpected part of our business that will grow very fast."

"This is a big breakthrough for the industry and certainly for us," Matza declares. "Some of our competitors that do HD dailies basically do their work twice. They'll go through the editorial and color-correction process in HD, and then from that HD master they'll create a digital cinema presentation, after which they'll go into digital intermediate and scan at a higher resolution. The digital preview screenings we did for *Van Helsing* included all the work that the DP and ourselves, and the director, put into creating the final product. It's a much better process than having it come from HD dailies off a telecine, which is not representative of how the film will look in the end.

"And when we're done with a digital intermediate, typically we're giving back an HD master, we're giving back an NTSC master, we're giving back everything the client and the studios need, not just for the theater, but also for all of their video releases and potentially digital cinema as well."

"Another trend we started that surprised us by its sudden growth is multiple negatives," Eicholz notes. "In a normal filmmaking process, you get a negative out of the camera, then you go to the interpositive—and you've lost one generation there—and you make another copy. You go to your negative—and lose another generation—and then you make prints, which is what's seen in theaters.

"What a lot of studios are having us do instead now is make multiple negatives, which we print on what's called SR stock. SR stock is a very strong stock that can be used to make multiple prints. And so you're taking the negative and you're making a first-generation print from that negative. What that means is that when audiences see release prints in the theaters, they can see the difference; they're crystal-clear. Some of the films we have done have six, seven, or eight copies that become the negative. This is why a room full of ARRI Laser recorders is a good thing" (Figure 7-4).

"This is something that the studios appreciate," Matza says, "they see the difference. Of course there's always pricing pressure, but the heads of postproduction at the studios are increasingly asking for multiple negatives."

"And one more item, one that's been in our business plan since day one, is to take the high-resolution digital masters and archive them in a computer system," Eicholz adds. "And then from that archival master, repurpose the content for our clients in any way they want—video, more film elements, digital cinema, whatever. We think that's a tremendous business. In fact, we think it could be bigger than everything else we've described."

Figure 7-4

ARRI Laser Film Recorder is the digital system EFILM uses for frame-by-frame transfers of high-resolution digital image data to motion-picture film.

Computer Power

At the heart of EFILM's data-intensive operations are computational, storage, and networking technologies made by Silicon Graphics (SGI), a company with a long history in making engines for high-resolution digital imaging and graphics.

"SGI has been our partner throughout the whole process," Matza explains. "Our technology team selected them because they're the only supercomputing company that can provide a robust development environment to handle our high-speed data requirements."

EFILM currently has three 16-processor SGI Onyx 3000 series visualization workstations (Figure 7-5) with SGI's high-definition, graphics-to-video (HD-GVO) option, a whopping 200 TB of SGI Total Performance 9400 storage, and the necessary high-speed connectivity to link the four color-timing suites with simultaneous streams of real-time 24-frame 2K images. And in addition to EFILM's four SGI Origin 300 servers, "All of our rooms, all of our machines, everything we do here, are all integrated on a CSFX multiplatform SAN [storage area network] environment, which means that we can share and move data very, very fast among the various components of our system," Matza adds.

Figure 7-5

Rows of Silicon Graphics Onyx 3000 computers and Origin 300 servers provide EFILM with the processing power and storage capacity needed for motion picture digital intermediate work.

"The graphics processors in the Onyx 3000 system allow us to display our images at up to 2K resolutions, and in real time. SGI systems—in combination with our proprietary Elab software and hardware—allow EFILM to design and configure multiple systems for multiple tasks. We're not trying to build a single suite, but rather multiple systems, each working on a different project with a number of parallel processes all happening at the same time. SGI met the spec we needed."

Matza and Eicholz both believe that 4K DIs will become common in a few years, and that increasing computer power and declining storage costs will make them economically feasible.

"My personal opinion about the advancements in computer technology—and the cost-performance—is that in the fairly near future, it's not going to be as costly to do it as it is now—for us and, thus, for our customers," Matza comments.

It's Worth It

Despite EFILM's increasing success in the DI arena, Matza advises that the industry needs to understand this latest digital cinema revolution for what it really is.

"Right now, everybody thinks that digital intermediates are a great new breakthrough process," he observes. "They're not. It's the same thing's been going on in video and digital mastering for the last 20 years. What *is* great about today's digital intermediates, however, is that they've gone on to the next step. Now you can *finesse* a film. The analogy I use is Photoshop. Yes, you can touch-up and color a photo in Photoshop, but you can also add textures, you can add layers. Now DPs have all of these new tools available to them to actually create new imagery in this finishing process as well as doing the most generic things such as color-timing."

"What's happening now is that digital intermediates are really in their second generation," Eicholz confirms. "And we're increasingly finding that the budget's already there for these extra services because studios have learned that these are things they want to do. We've done 36-plus digital intermediates at last count. There are many companies out there trying to do them; we have seven in-house currently, which are whole digital intermediates, where every single frame is in our system.

And really, our process is not that expensive when you consider the total cost of a film. Our process is on the back end; we're affecting every single frame, making the entire film look better. So spending some extra money to do that, it's worth every penny. And the studios have come to the same conclusion."

Sound Is Half
the Picture

Sound can be a paradox in the cinema. The visual portion of a movie tends to remain in the audience's conscious minds more prominently than the aural. But if the sound in a film is badly done, it will immediately draw unwanted attention to itself and disrupt the viewer's sense of immersion in the world the movie depicts. Under normal circumstances, visuals dominate the viewer's consciousness, but the sound tends to occupy the lower levels of awareness.

Given such an assertion, one might then conclude that sound is not all that important in motion pictures. This, however, couldn't be further from the truth. As director Jean Luc Goddard once noted, "Sound is 60 percent of cinema." And research by the Media Lab at the Massachusetts Institute of Technology found that audiences would tolerate a degraded video image far longer than they will a distorted soundtrack.

Although sight is often referred to as the "king of senses" and people unfortunately tend to take the gift of hearing for granted, we depend on sound far more than we realize. Music and sound effects—whether loud or subtly presented in a movie—can convey massive amounts of information and emotional context to reinforce visual imagery. Try renting the scariest horror movie you can find. Pop it in the DVD player, and mute the sound. How scary will it be? Frighteningly boring, perhaps, but not at all what the producers intended.

There are endemic differences between sight and hearing. Sound is ephemeral; it occurs, and—in an instant—is gone, often replaced by a succession of new sounds. It does not hold still and allow inspection, unlike visuals. Even a moving picture of a fast car chase shows you car and driver again and again, from many angles, against many backgrounds. Sound is literally invisible. These and other differences cause sound to hit you at a more personal and emotional level than visual information does.

Creative filmmakers well understand the differences between the auditory and the visual, and how it can be used to great effect. Sound is critical to the cinema experience, and is—in fact—where the first large-scale transition to digital technologies occurred in motion-picture production. It happened specifically in what's called postproduction, the editing and finishing process that takes place after a film has been photographed. In audio post, the sound portion of a movie (or TV show) is edited, enhanced, and re-recorded in preparation for its release. This is, by its nature, a labor-intensive discipline. As with their video counterparts, digital audio post tools have greatly improved the speed and creative flexibility by which motion-picture sound is crafted. Easily the single greatest event in the control of the production and presentation of audio for film is the digitization of the aural medium. It's a vital aspect of the digital cinema revolution that shouldn't be overlooked.

Reassociation

Walter Murch, the great film sound editor responsible for the landmark sound design of such films as *American Graffiti*, *Godfather II* and *Godfather III*, and *Apocalypse Now*, has a profound understanding of the role of film sound and the digital revolution. In an article in the October 1, 2000 *New York Times*, Murch wrote:

"Almost all of the technical advances in sound recording, manipulation, and exhibition since 1980 can be summed up in one word: digitization. The effect of digitization on the techniques and aesthetics of film sound is worth a book in itself, but it is enough to say at this point that it has continued forcefully in the direction of earlier techniques to liberate the shadow of sound and break up bottlenecks whenever they begin to form...to [let us] freely reassociate image and sound in different contexts and combinations."

The reassociation to which Murch refers is simply the ease with which we can now put any sound we want anywhere in a sound track, thus "reassociating" it with the visual appearing at that instant. Thus, we can record the sound of hitting a burlap sack full of rocks with an axe handle, and match it with footage of a 98 lb. actress punching a linebacker. Or we can play the digitally modified sound of an artillery cannon as an actor fires a handgun, and convey an impression of firepower.

In his *New York Times* article, Murch succinctly connects the historical role of film sound and its new position following the digital revolution. He points out how "our multimillion-year reflex of thinking of sound as a submissive causal shadow now works in the filmmaker's favor, and the audience is disposed to accept, within certain limits, these new juxtapositions as the truth."

The stage for the advances of digital audio was set in 1977, with the release of George Lucas' *Star Wars*. In the same way that *Star Wars* set new directions visually and thematically, the sound of the film became the benchmark against which all subsequent films were compared. The film took sound beyond conventional limits, presenting the popular imagination with an audio phantasmagoria that set a new standard for the way movies were designed and perceived.

Although the watershed soundtrack for *Star Wars* predated the introduction of digital audio production, it did inspire the demand for more and better sound in the cinema. It was made in an era when large, multitrack audio-mixing consoles from the record industry began to be adapted by motion-picture sound editors. Motion-picture sound mixing wasn't new, but these consoles allowed a greater number of individual effects, dialogue recordings, and music cues than ever before to each be assigned to separate channels for precise loudness settings, audio processing, and blending with other sounds. Lucasfilm's 1982 development of the nonlinear EditDroid further advanced the control editors had over both picture and sound. Additional early nonlinear editing technologies from companies such as Ediflex and CMX gradually began a transition process away from cutting film workprints and toward making "edit decision lists" based on videotape or hard-disk digital copies of source footage. The television and video worlds were the first to adapt these new tools, but the motion-picture industry eventually followed as the new digital sound- and picture-editing systems proved their flexibility and creative power.

MIDI Arrives

Long before telephone poles could be easily painted out of period dramas with digital effects software or computer-generated dinosaurs made to rampage across movie screens, motion-picture audio post—including music, dialogue, effects, and foley—had pretty much already become an all-digital process. (*Foley* refers to re-recording sound effects in sync with the picture.)

Since the late 1980s, when the job title "sound designer" became common, major motion pictures have benefited from the ability of creative talents to manipulate sound with computer technology. When converted into digital data, high-quality sound requires less information to be recorded, stored, edited, and reproduced than high-quality digital motion images do. It's one reason why nonlinear sound editing preceded nonlinear picture editing for feature films. Digital audio data provide all the benefits of any other kind of digital data, such as words in a word processor or calculations in a spreadsheet. Like any digital data, those representing audio are more portable, malleable, and copiable than their analog counterparts. They can be transmitted, edited, and reproduced much more easily, radically, and effectively than sound on an analog format like tape.

In the late 1980s, digital sound design and editing was originally performed on samplers, electronic instruments such as the Synclavier that were designed to let musicians use a keyboard to play back the digitally recorded sounds of anything from acoustic instruments to explosions. Movie sound editors, however, found that they could use samplers such as the E-mu Emulator to digitally record, edit, and sequence movie sounds much more quickly, cheaply, and extensively than with traditional means. The next technological innovation that further advanced the tools of digital audio post was the digital audio workstation, basically a "word processor" for sound.

Fairlight International, the manufacturer of one of the earliest digital audio workstations for music and audio production (during the mid 1980s), gained notoriety when composer Jan Hammer chose it to score his hot, electronic, jazz/fusion music for the TV show *Miami Vice*. What made the use of Fairlight revolutionary was that Hammer produced a new score weekly—on the fly—using the Fairlight as a digital music synthesizer and hard-disk audio recording/playback workstation. Hammer created, played, and mixed while watching the edited tape of each episode. In the process, he was helping define the way soundtracks could be created, edited, and posted digitally. Hammer's system also included a six-foot video screen and synchronizing gear for locking audio to video.

During the mid 1980s, companies such as Digidesign and Sonic Solutions developed improved hardware and software foundations for digital motion picture audio post. These foundations included computer cards for "off-the-shelf" (inexpensive) PCs that converted sound to digital data and back again, and the software needed to control the recording, playback, and editing of digital audio. In addition, pioneering programs such as Blank Software's Alchemy gave movie sound designers more, better, and faster ways to create never-before-heard sounds for film.

By this time as well, musicians and composers who used electronic musical instruments—synthesizers—were benefiting from the development of MIDI, the Musical Instrument Digital Interface. The MIDI specification laid out the basis for designing hardware and software that enabled computer control of electronic instruments. By the mid 1980s, a number of software developers were already offering sophisticated computer programs—MIDI sequencers—for desktop computers that would give composers complete digital control over enough MIDI instruments to simulate entire orchestras.

Toward the end of the decade came the digital nonlinear audio and video workstations developed by Avid Technology. In a few short years, the low cost and production flexibility offered by Avid editing systems quickly replaced traditional editing methods. From the beginning, the Avid system comprised elements made by other manufacturers, including an audio system built on hardware and software developed by Digidesign. This is significant because a large part of the acceptance of the Avid system in the film industry was because the Avid audio format was already compatible with Digidesign Pro Tools, and used by most of the sound designers, editors, and composers in the business. Avid eventually merged with Digidesign, but continues to operate as separate, cooperating divisions.

As the 1990s arrived, at least some of the music in most Hollywood films and television were created using computer-based, digital sounds; the music was produced using MIDI hardware and software, and the dialogue and effects went through at least some digital processing, if not being produced entirely in the digital domain.

Digital Advantages

Sound had for years been the last thing addressed in the filmmaking process, a hierarchy that went all the way from budgeting to implementation. Audio post schedules have traditionally been trapped between a film's planned release date and the frequent delays encountered in shooting the picture itself. Add to that the last-minute changes invariably called for by demanding directors, producers, censors, and so forth—changes that often must be made on the fly. An analog sound-track allows a certain number of such edits, but they are time-consuming and can eventually cause the physical medium to break down, further threatening schedules. Digital audio recording, however, presents an enormous leap of flexibility. With a digital audio edit system, not only do the standard computer techniques of cut, copy, and paste apply, but sounds can be merged, repitched, lengthened, sped up or slowed down, shortened, convolved, compressed, reversed, and otherwise altered in an almost infinite number of ways. And all of this manipulation is "nondestructive" to the original audio content; a simple keystroke can undo all changes to the audio content should producers have second thoughts about their sound edit. Not surprisingly, it took little time for the world's stressed-out composers, sound designers, and audio editors to see how digital flexibility made their lives easier and more creatively powerful.

Human ambition being what it is, however, the simple fact of making audio postproduction life easier proved to be just one of many benefits. The integration of digital technology into the postproduction arena also brought with it a new and enormous sonic palette. Sound designers now had the means to create sounds that were not only more realistic, but super realistic. Sound could not only represent reality, but also create a more-than-real soundtrack that could create an impact on the audience every bit as profound as the visuals (remember the laser blasts in the original *Star Wars*?). Once the technology was in place, filmmakers found themselves much freer to manipulate and heighten the impact of music, dialogue, and sound effects on the audience. And not only have the tools of audio post improved, so have the means by which that final, mixed, and crafted sound can be delivered to audiences.

Digital Soundtracks and Surround Sound

In the early 1990s, new digital soundtrack formats such as DTS, Sony's SDDS, and Dolby Digital began delivering vastly improved, multitrack audio reproduction to theater audiences worldwide. Using the tiny areas comprising the last available "real estate" on 35mm motion-picture film prints not already occupied by traditional analog soundtracks (the outer edges of the film, the space between film frames, or the space between the sprocket holes, respectively), these formats greatly advanced the quality of the sound audiences heard in theaters. Special readers, decoders, and additional speakers were required to implement these systems, but the clear, crisp, vibrant digital sound they produced made them a hit with audiences. Nearly all major films released today conform to all or most of these three formats. Add to this Lucasfilm's THX Cinema Certification program, which tests theaters for optimum sound and picture reproduction and which has already certified 4,000 auditoriums worldwide, and it's clear that audio is no longer an afterthought in the movie-making process.

Digital sound for theatrical movies is now the standard, and its benefits to the listener are multifold. Digital sound makes watching and listening to movies a much more immersive experience not only because of its clarity, but because the movie sound produced with the DTS, Sony SDDS, and Dolby Digital technologies also surround the audience. No longer limited to a single speaker or even a pair of stereo speakers, the typical movie soundtrack is now mixed in one of several different flavors of surround sound to come from as many as 10 distinct speakers. There are several surround sound formats.

Currently, the most common form of what is known as surround sound is called 5.1. It consists of six independent channels of monophonic audio coming through six speakers. In film, the center speaker usually carries the dialogue and occasionally part of the soundtrack. Front speakers on the left and right usually carry music and sound effects, and occasionally off-screen dialogue. A second pair of speakers (the surround speakers) is placed on the side walls slightly above the audience and toward the back. These carry effects and ambient sound. A subwoofer reproduces low-frequency effects such as (for example) dinosaur footsteps or monster spaceships flying close overhead.

Dolby Surround Pro-Logic was a successful surround system for home theater systems in the early 1990s; it became the standard for hi-fi VHS and analog broadcast television. Its successor, Dolby Digital (formerly known as Dolby AC-3), eventually overtook the earlier technology and is today the de facto standard, not only for cinema, but for DVD and HDTV. Dolby Digital's five main channels provide full frequency (20 Hz to 20,000 Hz) sound, and the sixth, low-frequency channel carries audio from 3 Hz to 120 Hz. Dolby Digital is backward-compatible with Dolby Surround Pro-Logic receivers, because the latter can "down-mix" Dolby Digital information and output the signal as stereo audio.

DTS Digital Surround is also a 5.1 channel format, and in addition to its use in movie theaters, it's an optional soundtrack on some DVDs. It is not a standard DVD format, and is not used by HDTV. DTS uses higher data rates than does Dolby Digital, causing some to believe it has better sound quality than Dolby Digital. It also uses more of a DVD's storage capacity than Dolby Digital.

DTS Extended Sound—a joint development program of Lucasfilm THX and Dolby Laboratories—is an "Extended Surround" sound format used in state-of-the-art movie theaters; a home theater version is called THX Surround EX. Lucasfilm THX licenses THX Surround EX for receivers and preamplifiers, and Dolby Laboratories licenses the technology under the name Dolby Digital EX for consumer home theater equipment. The Extended Surround formats add a surround back-channel, placed behind the audience, allowing more precise placement of both side and rear sounds. *Star Wars Episode I: The Phantom Menace* (1999) was the first movie to feature this format.

Both the THX Surround EX and DTS-ES formats encode the rear channel information into the left and right surround channels. Upon playback, the rear channel information is decoded ("derived") from the left and right surround channels. This encoding is referred to as matrix encoding and means that the rear channel is not a discrete channel. These formats are therefore technically considered 5.1-channel formats, and are sometimes referred to as "Dolby Digital 5.1 EX" or "DTS 5.1 ES." Although these formats are sometimes called "7.1 channel" formats, that's technically a misnomer.

In contrast to DTS-ES matrix, the DTS-ES format can optionally support a fully discrete rear channel. This true 6.1-channel format is thus called DTS-ES Discrete 6.1. Both matrix formats often use two rear speakers for greater coverage and impact. The Extended Surround formats are backward-compatible with their 5.1 counterparts.

Audio Impact

Given all the progress in printing digital soundtracks onto 35mm motion-picture release prints and then reproducing that sound so that it surrounds theater audiences, what sort of impact has it had on the cinema itself? Assuming that audiences are engrossed in viewing a high-quality film or digital projection of a movie (digital cinema servers are also capable of storing and reproducing multiple soundtrack and surround formats), the moment the audio starts coming from rear speakers and low frequencies are heard from a bank of big subwoofers vibrating the floor of the theater, the viewer is immersed into the middle of the action. And this need not apply only to loud sounds, such as bullets seeming to whiz past your head and burying themselves in the rear wall. Small sounds work differently as well. Crickets, footfalls, even just a breeze or a bit of echo are no longer emanating from a fixed point in front of you, but coming as they do in the natural world from in front, behind, left, right, above, or even below. Audiences are embedded within the action.

The overall effect here is subtle, even if specific effects are not. If you're watching *Saving Private Ryan*, you not only hear what you see, you also hear a barrage of battlefield horrors that you never see. They're behind you and off to either side and they bring you as close as a movie can to feeling present on the battlefield without actually being there. In a film as well crafted as *Saving Private Ryan*, this is not just a gimmick, but a powerful device for delivering, with yet further impact, one of the dominant messages of the film: the sheer evil of war.

Nor is surround sound the only technique in the digital audio toolbox. This is per Halberg, supervising sound editor for the 1999 Roland Emmerich film *The Patriot*, who invited antique firearm hobbyists to a desert rifle range where he digitally recorded everything from single shots to barrages of Colonial-era flintlock rifles. Halberg developed a palette of bullet whizzes ranging from musket balls to cannonballs. He also recorded the firing of modern weapons with much higher muzzle velocities. To create the movie sounds, he layered up to three or more whizzes together by placing them on separate digital audio tracks and playing them back simultaneously to create "ultra-realistic" sounds. Halberg was able to create dozens of combinations that yielded various combined sounds with noticeably different qualities. He even spoke of using certain sounds as leitmotifs, signaling a specific character or kind of event. He also spoke of mixing a certain scene where the main character (actor Mel Gibson) rescues his son. The scene had no music, and takes place some distance from the battle. Halberg used surround sound to separate the rescue scene from the main battle by placing no bullet sounds in the front speakers, but only in the back speakers. It was a subtle effect that worked remarkably well.

The Sound of Evil

Digital technology has proved itself as a great means by which to record, edit, store, and reproduce carefully crafted cinema sound, but the artistic use of sound in movies predates the digital cinema revolution by decades. A good example of this is the 1998 re-edit of director Orson Welles' 1957 *Touch of Evil*. Welles was absent from the final editing of the original film and was furious at the result. He sent Universal Studios a 58-page memo and nine pages of "sound notes," detailing how he wanted the film to be re-edited; Universal, however, implemented very few of Welles' suggestions.

Welles memo was, not surprisingly, ahead of its time; his film *Citizen Kane* (1938) remains to this day a landmark achievement in cinematic technique and artistry. What he had wanted for the sound in *Touch of Evil*, which was set in a Mexican border town, was not to have Henry Mancini's orchestral score dominate the film. Instead he wanted certain scenes set to background music as it would be heard drifting in from cantinas, radios, and jukeboxes. He wanted this music to sound exactly like it does when heard on the street in the real world.

"To get the effect we're looking for," Welles memo insisted, "it is absolutely vital that this music be played through a cheap horn in the alley outside the sound building. After this is recorded, it can be then loused up even further by the basic process of re-recording with a tinny exterior horn."

Welles distinguished this background music as not being the same thing as underscoring, which referred to Mancini's orchestral track. He wanted the background music to be, "Most sparingly used...musical color rather than movement; sustained washes of sound rather than...melodramatic or operatic scoring."

Needless to say, Welles was way ahead of his time. The distinctions he made in his *Touch of Evil* memo developed in the film world at large years later. Indeed, when multi-Oscar-winner Walter Murch was recruited to re-edit the 1998 version of the movie to conform to Welles' memo, he reports being surprised to read how Welles had described techniques generally thought to have evolved decades later, in such films as George Lucas' 1974 *American Graffiti*, on which Murch had been sound designer.

Ultimately, Murch used a bank of digital audio gear to re-do the entire track, making it as close to Welles' wishes as possible. Perhaps the crowning achievement of the film is its opening sequence: a three-minute continuous tracking shot that sets up the plot, characters, and flavor of the rest of the story. In the original release, not only did credits roll over this entire scene, but the only audible sound was Mancini's score.

Murch replaced this with the "background music" Welles asked for. He discovered an effects track obscured by Mancini's score on the opening shot that was never heard before. It had traffic, footsteps, goats, and other sounds. Murch digitally enhanced the track and saw that Welles had wanted "a complex montage of source cues." This he constructed using the Mancini soundtrack and a *Touch of Evil* music CD that Mancini had released in the 1980s, which included source music absent from the original film.

The result was that the rolling credits and Mancini underscoring Welles objected to were replaced by a mix of multisource rock, jazz, and Latin, fading in and out, along with a barrage of sound effects and dialogue central to the plot. It's a spectacular realistic introduction, foreshadowing the dark drama to come. Murch and his team re-edited the soundtrack of *Touch of Evil* to great effect using a brilliant director's original notes combined with the latest in digital audio post tools. The revised *Touch of Evil* is a good example of the power a soundtrack can have in enhancing cinematic drama. And with today's digital audio postproduction tools and surround sound exhibition technologies, the ability to craft and deliver the other "60 percent of the picture" has never been greater.

Rebel Without a Crew

"It's difficult to go up against tradition," Robert Rodriguez told an audience of filmmakers on the evening of July 17, 2003. The location was the Cary Grant Theater at Sony Pictures Studios, in Culver City, CA. The occasion was part of the Display L.A. conference, sponsored by the Hollywood Post Alliance and *Digital Cinema* magazine. Showing excerpts from his then-current films *Spy Kids 3D: Game Over* and *Once Upon a Time in Mexico*—which were projected in stunning fidelity using a Christie digital cinema projector outfitted with the Texas Instruments' 2K DLP Cinema imager—Rodriguez continued: "Creative people are slow to adapt to technology. But as [Pixar's] John Lasseter says, 'Technology challenges art and art challenges technology.' Digital HD really does push the boundaries of what you can do artistically; it frees you and makes filmmaking a much more vital experience."

Rodriguez, producer/director/screenwriter/DP/editor (and other *auteur* job titles too numerous to mention), had shot his previous three films using Sony HDW-F900 CineAlta digital HD cameras. He added that the only reason he'd ever shoot film again "would be for nostalgic reasons."

"Lose the viewfinder," he counseled, explaining that the immediacy of viewing HD monitors large and small during shooting means not having to wait for dailies to see if shots worked—especially when Steadicam is involved. This, he adds, enables him to work much more quickly, to see simultaneous live views from multiple cameras, and to get better performances from actors by showing them immediate HD playback of their work. Rodriguez also stated that HD was crucial to shooting *Spy Kids 3D* because "You can actively change the convergence and get the best 3D" whereas with filmed 3D, "you can't see what you're doing."

Rodriguez held forth confidently for more than two hours, comparing HD cinematography today to nonlinear editing a decade ago. ("Editors were afraid of Avids; now they won't go back to cutting on film.") He praised HD image capture. ("HD is the first time I get to see the movie the way it looked on the set.") And he addressed claims he's 'abandoning the art of film' ("Technology is not the art form, the art of capturing and manipulating images for visual storytelling is.") Then came the Ham Story. "I liken it to how long Hollywood takes to question tradition," Rodriguez explained.

"A young couple is making a ham," he began. "The wife is cutting the edges off the ham before she bakes it, and the husband asks, 'Why do that? It seems wasteful.'"

"'I don't know,' she answers, 'it's what my mother always did.'"

"So they ask the mother, 'How come you cut all the ends off the ham before you bake it?'"

"'Because that's the way grandmother always did it,' the mother says."

"So they ask the grandmother, 'Why did you always cut all the ends off the ham before you baked it?'"

"And the grandmother replies, 'Well, that's the only way I could get the ham to fit in the pan I had.'"

"A lot of the things we do, we do them because they've always been done that way," Rodriguez concluded. "But once you take the leap and start using that 'bigger pan' of HD, you realize 'Hey, we can do everything differently now.' And then you question the entire system, and that's a good thing."

Lab Rat to Filmmaker

A bit of a ham himself, Rodriguez is quite outspoken about the advantages of digital HD as an alternative to—perhaps a permanent replacement for—35mm film for cinematography (Figure 9-1).

"Digital HD isn't the *future* of filmmaking," he explains, "it's the *now*. It has quietly arrived and it's here to stay. I don't think everyone realizes this yet. But as soon as you bring an HD camera to your set, it's over; you'll never go back to film. HD offers up so many new possibilities; you can cre-

Figure 9-1

Director Robert Rodriquez on location using a Sony HDW-F900 HDCAM on a Steadicam rig shooting *Once Upon a Time in Mexico.*
(*Photo by Rico Torres.*)

ate a new movie language with it. Now we won't just be living off the inheritance of tricks learned from decades of past filmmakers. Digital HD enables us to come up with something new."

Like the action heroes in his films, Rodriguez is a rebel. The Texas-born filmmaker burst onto the movie scene in 1992 at the age of 24 with *El Mariachi*, a $7,000 feature for which *he* was the entire crew. He financed the movie by working as what he calls a "human lab rat" for a drug company during his summer vacation from the University of Texas at Austin.

"The drug they were testing was a speed-healing drug," he writes in his 1995 book, *Rebel Without a Crew* (Plume/Penguin, ISBN 0-452-27187-8). "In order to test it *they had to wound you.* Now I've got two tiny football-shaped scars to remind me of how I used to finance my films."

Columbia Pictures bought *El Mariachi*, funded its blow-up to 35mm, and signed him to direct a sequel, *Desperado*, for $3.1 million. Rodriguez then went on to score continued triumphs with such fast-action films as

From Dusk Till Dawn (1996), *The Faculty* (1998), and his family-friendly blockbuster *Spy Kids* (2000).

Rodriguez not only writes and directs his movies, he also prefers to serve as production designer, director of photography, and editor. "I try to do all the key jobs because it's actually easier that way," he explains. "You don't have to have a meeting with somebody every time you get an idea, which is all the time." Knowing of his hands-on interest in production technology, George Lucas showed him early digital HD footage of *Star Wars Episode II* during a January 2000 *Spy Kids* sound-mixing session at Skywalker Ranch. This prompted Rodriguez to do a side-by-side test of 35mm film and Sony's HDW-F900 HDCAM CineAlta 24P camera system.

"When I screened the film-out I was shocked to see how bad film looked compared to HD-originated film," Rodriguez recalls. "The studio [Miramax] couldn't understand why anyone would shoot film after seeing those tests. So they gave me their blessing to try it. I was so convinced after I did that test I went out and bought two F900s for myself."

Rodriguez then embarked on shooting *Once Upon a Time in Mexico* (2002; the third in his *El Mariachi* series) and *Spy Kids 2: The Island of Lost Dreams* (also 2002). The following interview was conducted jointly by phone in July of that year by the author of this book and Bob Zahn, President of Broadcast Video Rentals, in New York.

Why shoot in digital HD?

I was always disappointed by the limitations of film. I don't find it to be creatively conducive. It's really limiting in how fast you can move. With HD, when I'm lighting, I'm looking at the monitor, I'm looking at the set, knowing how it's going to translate. I check the monitor, and we just move, we don't stay there forever wondering, so it's very fast.

Can you elaborate on limitations of film?

There's a lot of technical hang-ups to film. I DP my own pictures, pick the lenses, operate all the cameras and Steadicams, and pick all the angles. Usually I had someone just keeping track of the exposures because film was so unpredictable. I didn't want to have to be messing around with all the *f*-stops and wondering, 'Oh gosh, how's this film gonna behave with this lighting? Is it going to cook something or underexpose something too much?' It was always so unpredictable.

HD is very freeing and is more like going back to the basics of filmmaking, where it's fun again. It's just so much easier to shoot in HD. Since I'm my own DP, I'm able to see exactly what I'm getting on the monitor, and I'm able to be much edgier with the lighting. When I was shooting Johnny Depp in Mexico, I would sometimes just use the bounce card and a little piece of tinfoil bounced into his eye and I knew at what level to do it because I'm watching it on the monitor, and then we moved on and kept going.

There's no guesswork or waiting for dailies. We moved a lot faster and it was a lot more satisfying. I also edit my movies, and it felt like HD is like the difference between cutting on film and cutting on an Avid; it was that big a change in the creative process.

What other things did you like about digital HD?

The F900s were so much lighter on the Steadicam, which is good on my back. And since you could take the tape, play it back, and see what you were getting, you knew when you nailed it; then you could move on. The freedom that HD gives you saves so much time, money, and headaches. For that reason alone, you should just ditch film.

I also do a lot of my own production design, so while I'm designing the set, I'm already thinking about how I'm going to light it. But when I get the film back, I'm always disappointed because it never looks like it did when we were shooting on the set. HD turned that around. HD was the first time I saw that what I was getting was what I had seen on the set. With film, it's always downhill from the moment you walk on the set until you finally see your movie released.

Everything we do now ends up as work on the screen. Every color we paint isn't all turning gray like it does with film, the color isn't sucked out of it; we don't have this extreme amount of contrast that film does now these days because of the way they process it.

How did shooting digital HD influence your work with actors?

Since I didn't have to cut, my takes would go longer. That really helps when you're in the flow with an actor, to just keep doing the take over and over till you get it right. You don't want to cut.

I found that in shooting children for *Spy Kids*, I had to let the camera run a lot so I could get the best takes. And the time spent running out of film right when they were getting warmed up was just brutal. I think any filmmaker who compares film and digital on-set will suddenly look at their film camera like it's a lead brick or an old vinyl LP record, and realize, 'There's recordable CD now; what am I doing with this vinyl record that requires me to change sides?'

Did digital HD save you time and/or money?

It saved a lot of time, which always saves money. I shot *Once Upon A Time In Mexico* before *Spy Kids 2* to learn what the strengths and limitations of digital HD were. I never could have made *Mexico* on film. It would have been too much work—too much trouble—to get that kind of movie onto film. The only reason I did it was because of the possibilities of digital HD; we shot it in the same amount of time that we shot *Desperado*, seven weeks. But *Mexico* is a much bigger movie. We saved so much time each day because we knew what we were getting. That alone, to see what you're doing, just lets you move faster. That's just how it goes.

But 'cheaper' isn't why I use HD. I'd use it even if it was more expensive. Do you think everyone who's cutting on an Avid is doing that because it's cheaper than cutting on film? No, it's 10 times more expensive. But the process is 100 times better. Of course you're going to eat the cost because it's a better process. With HD you're actually saving money, *and* it's a better process.

Tell us more about that process.

HD has changed the creative process. People will be surprised how enormously entertaining these movies are, and it's because of that changed filmmaking process. You can actually make a much better movie by shooting on HD than you can on film just because of the process, and people don't realize that yet.

Most actors don't like to see themselves after a take, but I would drag them over to the HD monitors so they could see. It's exciting, like being at a premiere while you're there on the set. A lot of times with actors you're pushing them in a certain direction; they always have doubts: "Are you sure that was good?" Now I can drag them over to the monitor and say, 'Here, look for yourself.' And they were always impressed. I think everyone found it helpful.

Everyone watching there on the set could see what they were doing and they would improve upon it. That never happens on film. And everyone was surprised by how good the image was on the HD monitors. You can make each moment count much more with HD. It really helped with the kids to be able to show them exactly what they had done, and to tell them what to do next.

Film is like painting on a canvas in the dark; you don't get to see what you did until the next day when the dailies arrive. It's as if the lights turn on the next day. You go home each day after shooting film thinking, 'I wonder if I was even hitting the mark? I wonder if I was even on the canvas?' You have no clue, and it's a ridiculous way to work. So for that reason alone, there's no point in shooting film unless you like to guess. With HD, you finally can see what you're doing, and you can do much better work. The HD process alone is the reward.

Do you consider yourself an artist?

I'm a filmmaker; I don't really like just to direct in and of itself. That's what I love about HD; it makes it possible to do all the jobs yourself if you want, because it's that easy to be creative. It may sound kind of crazy, but it really frees you to be a shoot-from-the-hip filmmaker telling the stories you want very easily and quickly. Again, it's just like the Avid. You don't need to have several people there to take your film and hang it in bins and then go find it. It's all there automatically so you can just sit there and cut your whole movie if you want.

Actually, for me, the most fun is editing. But to edit something really cool, you've got to be the guy writing it, getting the right shots, making sure it looks nice, and getting the performances. You've got to kind of do everything to get to that point where you can then be the editor who gets to put it all together and see the end results.

Visual effects were an important part of Spy Kids; did digital HD help in that regard when you shot the sequel?

Any time you do an optical in film, you lose another generation. It's much easier to pull a matte digitally [in HD] than with film. With HD, it pulls as quick as it would from regular video. It's instant. When we filmed the first *Spy Kids*, we wanted to get the cleanest images for pulling mattes, so we shot slower film stock, which meant a lot more lights on the green screen, which is a lot more money. And I still wasn't satisfied with how it looked. I'd visit the effects guys and they were having so much trouble pulling mattes, taking the grain away, pulling the matte, putting the grain back in. I mean it was just ridiculous. Film is so archaic, it's just not worth it anymore. We did a lot of green screen on *Spy Kids 2* and it was so much easier this time.

Explosions, stunts, and slow-motion effects are key to your movies; how did you overcrank 24p?

I actually got the slow-motion idea from E-FILM. They had done a test for me and it worked really well, to use the 60i. Now my effects company, Troublemaker Digital Studios, does the slow-motion conversion. They make it 30 frames per second or 60 frames per second, which looks really rich and good. Now when I want it slow motion, I shoot at 60i.

Originally, I wasn't sure if the slow-motion tests would work, so for the first two days, we had a 35mm camera on set; it looked like a dinosaur. That was our crash camera. And then right after the second day the tests came back, and the slow motion was satisfactory, we got rid of the film camera.

But I did get to record a few explosions with both the film camera and HD, and HD held up much better than film. The explosion wasn't as rich on film as it was on the HD. I was really surprised about that because it was in a dark room and I thought the stop difference would just blow out completely and the HD wouldn't retain the detail of the film. But it was completely the opposite. It looked horrible on film. It's so contrasty that it doesn't even look like an explosion.

If you set up your HD cameras upright, you can really get some incredible things. I always do my own camerawork, and I like the freedom of zooms [zoom lenses]. On film, the image always suffers when you would use the zoom, but with HD, I could get all my different lens sizes within one take, because I'm also the editor of the film. So if I'm sitting there shooting, I'm not going to shoot a whole take wide and then a whole take tight. I do my wides and my tight coverage and medium coverage within the same take because I already know how I'm going to cut it together.

Johnny Depp kept saying, "What I love the most is I don't know if you're full-body on me or right on my eyeball, and I love that because I know I can't fake it in any way. I never know where you are with that take, because it's not like the old days where it's 'Oh, you've got a 50mm on me? I'll act a certain way.' Or, 'It's a 250mm? Okay, I'll be more subtle.'" You can't do that now.

Did you do any special processing to the 24p HDCAM signal; was it captured on other media, other than the recorder inside the camera?

The only time we did that is when we used the 950. We used the 950 a few times to have a more mobile camera, but even then I think we just recorded to a regular deck. We didn't do anything like George [Lucas] might have done—recording to hard drive. We didn't get that fancy yet. But hopefully the next step, when we get to that, would be going 10-bit with the recording medium.

Were there any special challenges in shooting in digital HD?

You should always check your back focus in temperature changes. That's just something that you have to be aware of. You have to check the monitors to see if something's going soft.

Was audio recorded on the HDCAM or on a separate system?

We did it both ways. We had it going to the HDCAM, but the second system was what we resynched everything to later using a hard-drive recorder.

Are you happy with the quality of your HD images?

I've seen the film-out, and it's incredible, it's very rich. I took advantage of what HD offers when I did the set design. I lit them [the sets] very rich, and the colors are like watching a Technicolor movie. It's so rich and beautiful to watch that I think it's going to wake a lot of people up. People aren't going to believe how good these images are. They'll be pinching themselves.

George Lucas said he'd never shoot another film on film again. Do you feel the same way?

Absolutely. I don't even shoot still pictures on film anymore. I buried my Nikon. I have kids, and film is just ridiculous for shooting them. You know when you've got the shot so you can walk away knowing that you've captured the moment.

What advice would you offer to other directors considering using digital 24p HD instead of film?

Digital requires a learning curve. You have to get in there and use it. It's hard to find anyone to teach you because everyone has a different way of doing it and not all of them work. You're not going to find that out until you're a few weeks into filming. You've got to be aware that that always happens with a new medium.

But the rewards of digital HD are just so great. And since you've got a monitor there, there's not a whole lot you can't fix. It's really worth the challenge. There's a lot of things you can do to the camera. If the image doesn't look right, it probably isn't right. So I would just tell them to get into it because digital is the future. It's really where things will be going.

What would you say to filmmakers who are intimidated by digital HD?

They shouldn't. Editors were afraid when Avids were being introduced. They said, 'We have to change our whole way of working and learn all this computer stuff!' But now everyone edits on them. With digital HD, they've just got to get over it and learn and embrace the new technology so they're not dinosaurs. HD is their friend. They don't have to be guessing anymore. They can really do much better, much edgier, much more exciting lighting because they are able to see the result instead of waiting for that dreaded daily report to come in. They can find out right then and there if it worked.

HD is in its infancy. This is the worst it will look. I can't wait for the next generation.

Indy Dreams, Indy Realities

Has the digital cinema revolution really arrived in terms of independent filmmaking? After all, now anyone with the passion to make a movie and a credit card to buy or rent the new generation of affordable digital production gear can be a filmmaker, right? And it's only appropriate that indys should be the ones to lead the charge. After all, independent filmmaking is where the movies began; how else would one describe Edison, the Lumières, Méliès, and the other cinema pioneers? No longer will giant conglomerates be the only ones to decide what movies will get made. We are now in an age of indy filmmaking nirvana, right?

Not necessarily.

"The marketing pitch by equipment manufacturers that says: 'Pick up a camera, anyone can make a movie' just isn't reality," warns John Manulis, CEO/Producer at Visionbox Media Group. "The fact is that 'just anyone' *can't* just make a movie. You need expertise and experience or you have to do an awful lot of testing and workshopping on your own to acquire such experience. This doesn't diminish my enthusiasm for what digital formats offer, it's just that there's...more to filmmaking than getting a digital camera."

Manulis knows digital filmmaking. As a co-founder of the "creative incubator" that is Visionbox he's had a direct hand in guiding many high-quality independent digital films through to completion. These films include such critically acclaimed titles as *Tortilla Soup* (2001), *Falling Like This* (2000), *Teddy Bears' Picnic* (2002), and *Charlotte Sometimes* (2002). The first time I spoke to Manulis was in late 2000. The result of that conversation follows, below, in this chapter. The next time I spoke to him was in early 2004. And although his enthusiasm for the digital indy filmmaking revolution was undampened, it was also tempered by the experience of the intervening three years. Yes, talented filmmakers do have greater access to the tools of production than ever before. But many other factors in the long-established business of the movies aren't quite so simple or easy that they can be changed overnight. Sundance notwithstanding, the excitement of totally democratized, digital indy filmmaking dreams is often rudely awakened by the harsh world of digital indy filmmaking realities.

"Another thing that we haven't seen—and it was something that a lot of people were predicting—is a change in the hegemony of the theatrical distribution model, or the studios' involvement in that," Manulis continues. "However little it costs to make a film on a digital format you still have to get it out to an audience and you still have to buy the marketing that gets the public to know you're there. I question whether in fact digital cinema's going to change that situation. I don't know that we've found the model that proves that your visionary, low-cost independent digital film will be successful just because it's been made.

"On an independent level, digital production is everywhere," he acknowledges. "But in a mass-market, commercial way I think we're going to see digital presentation become part of the fabric of the entertainment world faster than digital production will. The economic incentives of digital presentation are so much stronger for the studios—as distribution entities—to take advantage of, especially now that Digital Cinema Initiatives is getting its act together."

Digital Cinema Initiatives (DCI) is the consortium of seven studios devising technical recommendations for digital production, post, and exhibition. Working with the Digital Cinema Laboratory—an organization established by the University of Southern California's Entertainment Technology Center—DCI has been taking a step-by-step approach to figuring out how much of the promise of the digital cinema age—especially as it relates to exhibition—is fact and how much is fiction. They're not alone. As the transition proceeds from an analog, film-based cinema to an improved digital one, there are many lessons being learned.

"In the course of making our own films we became aware that there's a whole integrated suite of services missing out there in the independent world," Manulis reveals. "People asked us to step in and help them with their projects, so our service business evolved, and with it the formation of Visionbox Media Group with postproduction partner Chris Miller. This includes physical production, the packaging, finance-consulting, sales, marketing, distribution-consulting, and postproduction.

In addition to expanding its consulting services, Visionbox also gained valuable lessons regarding indy film financing.

"Digital is still something people would rather not do unless it's an absolute necessity for getting a movie made. In dealing with both international and domestic markets we found it was a bit of a miscalculation to think that there was going to be a big appetite for the under-$100,000 kind of product. We learned that it's just as hard to make a movie for $100,000 as it is for a million or ten million dollars. At a certain point everyone realized that to make a movie that was deliverable and salable to world markets, there's a lot of time and expertise that has to go into it, particularly when you're working with digital."

Sobering words, perhaps, for low-budget digital indy filmmakers, but Manulis' comments are based on experience.

Late 2000

"Novels are different from poetry," Manulis observed during our first conversation in late 2000. "Sixteen millimeter is different from 35. Black & white is different from color. It wasn't always cost that made certain filmmakers work in black & white, or that made others shoot in 16. And the same thing is true about shooting on Mini-DV versus shooting on HD. I think they're simply tools for different types of storytelling."

Manulis knows. As the Co-Founder and CEO of Visionbox Pictures, a digital production company based in Culver City CA, he and partner Michael Kastenbaum produced writer/director Harry Shearer's theatrical feature *Teddy Bears' Picnic* (2002), which was photographed with Panasonic AJ-PD900WA 480p DVCPRO50 camcorders. (Panasonic has since replaced its digital cinematography camera with the 720p Varicam; see Chapter 5.)

"Image quality is a significant part of everything we do," Manulis explains. "We're a full, traditional distribution outlet, and our pictures lend themselves to other platforms because of our ideas and the ways that we conceive of them. But fundamentally we're playing into theatrical and ancillaries across the world, so we have to start with great image quality.

"Michael and I are passionate advocates of digital technology and of pioneering a creatively liberating production process," he continues. "Digital is creatively empowering and it enables you to find a way to finance projects that people might otherwise talk their way out of because they're basically risky. Digital also enables you to retain ownership if you put your deal together right. From both a creative and a business level it's pretty intriguing.

"That being said," Manulis adds, "I'm still having to 'sell' people heavily on digital. At the moment those people are financiers, distributors, and filmmakers. More and more of them—particularly filmmakers—are having some level of exposure to digital, but they still have the same questions, and you have to explain the same things, and show the same tests. That's just part of what you take on when you jump into something like this early."

The Right Medium

"Digital's early adopters really took advantage of its portability," Manulis recalls. "You had a lot of 'run-and-gun' shoots, things shot with a moving-camera in very aggressive *vérité* fashion and with very little lighting. But if you're not into that style of shooting, right away you begin to think that it goes hand-in-hand with 'digital.' You can tolerate that style of shooting once or twice, but you certainly don't want a steady diet of it.

"So part of our convincing process has been showing people better image quality, showing them that you can, in fact, shoot digitally on dollies, you can shoot on sticks, you can shoot in a more formal manner, you can light more carefully. We just won the Best Film and Best Cinematography Awards at the Hamptons International Film Festival with a digital picture— *Falling Like This*—which was shot in straight DV."

Manulis relates that it was the lab tests that convinced Visionbox that Panasonic 480p DVCPRO50 was the right medium for shooting two of their most recent features.

"We did a couple of different tests of digital converted to 35mm," he recalls, "and we've just finished one that took the same footage and ran it through five color-correction labs and five film-output labs in a sort of full-permutational test. Before that I'd done the same test with different camera systems going to film.

"Choosing 480p wasn't that difficult a decision by that point," he continues, "it was an easy test to look at. I've shown it to a lot people, and no one has yet disagreed. The only thing that might have contended [with our choice of 480p] was using the full digital HD system, but a lot of us felt it was *too* clean and crisp. That format is good for George Lucas because he's making an event movie, a big spectacle focused on action. But a lot of people felt that for a 'human' movie, full digital HD would actually be too intense. When you're dealing with human interactions there's a certain degree of humanity that you want in the images so audiences can identify with them. Plus, the 480p DVCPRO50 format has significant advantages in terms of the size and maneuverability of the equipment and the cost of postproduction to high-def."

Maximizing Time

Teddy Bears' Picnic is one of the "human" movies Manulis refers to. Written and directed by Harry Shearer (*This is Spinal Tap, The Simpsons*), the film is a comic look into the world of high-powered corporate retreats. The feature was shot by cinematographer Jaime Reynoso. According to Shearer, 480p "greatly enhanced" the production process of *Teddy Bears' Picnic* and provided several advantages.

"The obvious one is cost," Shearer says. "Tape is a lot cheaper than film, and you can keep rolling to get material, which, while not needed at the time, comes in very handy in the editing suite."

"Digital accomplished most of what I expected, which was to speed up the production process, and eliminate many of the delays usually attendant upon a film shoot," Shearer adds. "The flow of scenes was never interrupted for reloads. My main concern was maximizing the limited amount of acting time I had available to get the maximum amount of useable performance choices."

Shearer also adds that 480p provided "The ability to avoid certain technical problems during shooting by having access to digital post technology." The script called for green-screen shots, and as Manulis explains, "We were able to facilitate that compositing process in our native medium."

As to whether Shearer would use a digital camera again for a film, he responds, "Absolutely."

Less Was More

Another "human" movie Manulis produced digitally (although not a Visionbox production) was *Tortilla Soup* (2001), which is based on Ang Lee's film *Eat Drink Man Woman* (1994). *Tortilla Soup* is an emotional comedy about a Los Angeles-based single father, his three daughters, and the Mexican cuisine that binds them together. A Presentation of Samuel Goldwyn Films and Starz Media, *Tortilla Soup* was directed by Maria Ripoll; the DP was Xavier Perez-Grobet, who shot *Sexo, Pudor y Lágrimas* (*Sex, Shame, and Tears*, 1999), the most successful movie in Mexican film history.

"We were looking for flexibility with lighting and, given that this is a very performance-oriented piece, for ways to free the actors from the traditional film constraints of limited footage," Manulis says.

"Over half the movie takes place in one location, so most important was the ability to maximize the amount of rehearsal and performance time for the director and actors. Panasonic's 480p DVCPRO50 allowed us to shoot a lot of footage, using a B camera consistently to roam the action, cover extra angles, and capture behavioral details. Digital also provided the aesthetic attraction of being able to shoot in a practical location with a smaller package, less light, less heat, and fewer people, and the practical attraction of achieving cost-savings on a project with difficult pre-sale capacity" (Figure 10-1).

Figure 10-1

An all-star cast assembles at the dinner table for director Maria Ripoll's 2001 digital film *Tortilla Soup.* Director of Photography Xavier Perez-Grobet adjusts the Panasonic AJ-PD900WA 480p DVCPRO50 camera, at left.

Time Away

"We are synching dailies and transferring to 3/4-inch to digitize into an Avid for editing," Manulis explains, regarding post for *Teddy Bears' Picnic*. "We will use an EDL list to up-rez from the 480p masters to 1080i and go online for color-correction, compositing, titles, etcetera, and then output a Digital Betacam tape for immediate use and festival consideration. Later, we output hi-def files for the creation of a 35mm negative. We had a 16-week postproduction schedule, not including the transfer to film.

"We followed virtually the same post process for *Tortilla Soup*, with the exception that we involved our postproduction partner, Digital Difference (Santa Monica CA), from early pre-production. We worked with Kevin Hearst and Chris Miller there to expand our ability to bring a consolidated suite of high-quality, low-cost digital post services to independent filmmakers. This technology is changing every day, and in a constant state of flux. We want to get different systems to work together, and to evolve ways to make them more friendly and cost-effective for independent filmmakers for a soup-to-nuts postproduction solution. Both films have been benefiting from a combined research and beta process in post, which is providing a major test of the output-to-film processes from numerous labs and formats.

"You can do a vast part of your postproduction and take stuff a long way through the process on Apple Final Cut Pro and intense Mac G4 set-ups," Manulis adds. "This enables something really valuable, which is 'time away'; often these machines are not only used in facilities, but are *owned* outright by filmmakers and set up in their living rooms. On *Falling Like This*, writer/director Dani Minnick, producer Lulu Zezza, and editor/DP Alessandro Zezza were able to take six months more in postproduction because they worked at home and weren't burning money every week running systems and paying editors in a facility. As a result the discovery process was much better. They had the ability to not feel pressured, to step away, come back two weeks later, and judge their work with fresh eyes. That's a gift in this process, and very few people in Hollywood or elsewhere get it. It's a great thing what this new digital filmmaking technology can avail to people."

Tools and Talent

"We're not closing the door on film, but we are encouraging filmmakers to take advantage of our experience with the flexibility of digital, and to apply themselves creatively to the advantages and differences of the medium," Manulis says. "We are obliged to stay on top of all the new equipment and be responsive to filmmakers' own aesthetic and qualitative judgments.

"The Panasonic 480p systems have performed very well for us," he adds. "We rented camera packages from Birns and Sawyer and Plus8Video. The cameras were rigged for focus pulling and filters, and there are nice advantages because you have that [attach-on] LCD viewfinder that pops out from the side, which lets the director or DP move alongside a camera and see what's going on while the operator's in the viewfinder. If you're moving with the camera the LCD is a great tool and a really handy item. [Panasonic has since introduced 720p HD digital cinematography cameras that have eclipsed the capabilities of its 480p models (see Chapter 5).]

"Having the tools does not make everyone a good storyteller," Manulis cautions. "A good DP is still a good DP: someone who knows how to work with light. Once they make the adjustment to what this particular system's quirks are, they're still going to be better at creating images than someone who hasn't trained as a DP.

"I'm torn about digital because great films have been made on *film*, and with very limited amounts of it, so who knows what's 'better' in certain circumstances? But I definitely did find that digital's advantage of enabling you to have fewer people around and get the sense that the set is *more* committed to the actors, the director, the performances, and the techs— brings greater potential to have their needs and rhythms carry the day. I wouldn't say that there was so much time that we did endless takes once we were working in this way. But we certainly were able to run, and to give them more opportunities. We were certainly able to shoot rehearsals in a different way. We ran two cameras more aggressively than we ever would have been able to with 35mm.

"I think we'll still see the same ratio of good product to bad," Manulis concludes, predicting digital's future impact on moviemaking. "But if new technology helps creative ideas to be more viable either because of lower economic thresholds or because of self-empowerment—people being able to have a lot of their own systems—I think that's better and it can only lead to exciting work."

A Whole New Ballgame

Hollywood has reigned supreme as the Western World's filmmaking capital for nearly a century. Digital filmmaking has the potential to change that. If movie making is democratized by affordable production tools, who's to say it must continue to be centralized in a specific geographical area?

"Go West," has traditionally been good advice for filmmakers. The high costs of making movies tended to concentrate that industry in Southern California with its year 'round sunshine. Runaway production has, more recently, made Canada, Britain, Australia, Mexico, and Eastern Europe attractive alternatives to the rising price of shooting in Los Angeles, although Hollywood still calls the shots and controls the distribution pipeline.

Now, however, the relatively reduced costs of digital cinema tools makes regional filmmaking in the U.S. (or anywhere else) far more feasible. Whether or not the Internet or other new media pipelines of the digital age also democratize film distribution remains to be seen. What is evident is the high level of production quality and expertise seen in new regional filmmaking centers. Cincinnati's Mike Caporale is a good example.

Leveraging Experience

Originally a painter and photographer, Caporale transitioned to 35mm film during the 1980s, assembling a complete production package for shooting commercials, industrials, and other assignments. In the 1990s he opened Finis, one of the Midwest's first digital post facilities. Next came Caporale

Studios to serve a variety of clients needing national spots, corporate videos, and video news releases. Through it all, however, Caporale's passion for filmmaking continued unabated. He leveraged his commercial production and digital post experience toward jobs as second-unit director on such independent digital features as *Welcome Home Roxy Carmichael* (1990, starring Winona Ryder) and as DP on documentaries spotlighting the homeless and the late bluegrass master Bill Monroe. An invitation to organize a seminar at the 2000 Southeastern Media Film Conference, however, proved to be a turning point in his career.

"It was a great two-day seminar," Caporale explains. "It was titled 'The Practical Application of Videography for Die-Hard Filmmakers,' and it was held at the Screen Gems Studios in Wilmington NC. I had enlisted the help of Panasonic to support it and they sent several engineers. They brought in their HD projector, their HD camera, and monitors. On day one we basically gave an overview of all the choices in digital production, and then screened material on the HD projector so everyone got to see literally every format up- and down-converted to HD and projected in HD, and that included both video and film. Then on the next day we went into the screening room at Screen Gems and we projected, with Sony's cooperation, footage that had almost every format of digital video converted to film" (Figure 11-1).

This led to Caporale eventually meeting producers Allen Serkin, Jason Davis, and Damien Lahey, who offered him the DP job on their fea-

Figure 11-1

Director of Photography Mike Caporale with hand-held Panasonic AJ-PD900WA digital camera in the dugout shooting actor Rob Piper (center).

ture *Ball of Wax*, a satirical look at baseball. Having seen the digital camera and projector demos, the producers and director/screenwriter Dan Kraus agreed they wanted to shoot the film using what was, at the time, leading-edge technology. That was Panasonic's AJ-PD900WA DVCPRO-50 camcorder, which produced images in "480p," or 480-line video, progressively scanned. Caporale was already familiar with the camera from having used it on many Caporale Studios assignments.

Advantages

"We had a three-week shoot," Caporale recalls, "and I really put the camera through its paces, encountering every conceivable lighting condition. The camera was able to capture amazing nuances of the event.

"Because of the camcorder's light weight and maneuverability, we were able to get some really extreme shots," he adds, "including an actor literally flipping the camera in the air and recording the spiraling POV, shooting up through the steel grid of a shopping cart, and holding the camcorder at waist-level and whipping from left to right to capture a frenetic argument between a husband and wife. The camcorder itself could be used as a unique means of expression because it's more mobile."

Caporale explains that the budget for *Ball of Wax* was only $35,000, which included a full crew, actors, baseball teams, full uniforms for the home team, location photography in various North Carolina sports venues, and night scenes.

"It would have been impossible to shoot it on 35mm for that price," he says. "Speaking in round numbers, if you figure $1,000 for an 11-minute roll of 35mm for shooting, processing, and some form of dailies—this could cost more or less, depending on how you cut your deals—shooting an hour's worth of film at a ten-to-one ratio is $60,000; double that for two hours. We spent probably $1,000 on tape.

Digital is Film, Film is Digital

Ball of Wax went on to win Best Digital Feature film at the Cinequest Film Festival in 2002. Panasonic had upped the ante for digital filmmakers the previous year by introducing its 720p Varicam, the world's first variable-frame rate progressive-scan HD cinema camera (see Chapter 6).

"I bought one because it was the hottest damn thing on the planet," Caporale recalls. "It's truly phenomenal. After working with that camera it was hard to look back at anything else."

Panasonic then introduced CineGamma, software that permits its Varicam to more closely match the latitude of film stocks. Caporale sold off the last of his film equipment.

"If you were working in film you'd have to do all your opticals for your titles, dissolves, and effects," Caporale observes. "You'd have to get all your audio converted to opticals. You'd also have to go through the process of answer prints, which is absolutely maddening. You never get an answer print right the first time, and it goes back again and again, and you're doing all this color correction, and you don't have the ability to do the type of color correction that you can do in a telecine suite. With digital color correction you can create moods electronically at the touch of a button. And depending on how you handle it—if you do it very selectively and carefully—you can digitally process your film for a few thousand dollars, and really have it what we call 'spanking' (Figure 11-2).

"It's a funny thing," he recalls, "when you shoot with the Varicam at 24 frames it looks like film. I'm happy with good-looking video when it's printed to film, because once it hits film it is film, and it looks like film. But that 720p variable-frame-rate camera, when it's shot at 24, even before it hits film, even when you're watching it on video, it looks like film. It's about a 600 ASA camera...very little is necessary to make it look like film. And

Figure 11-2

Director of Photography Mike Caporale mounts Panasonic AJ-PD900WA on the base of a shopping cart while shooting *Ball of Wax*.

because of its 11-stop dynamic range, the CineGamma curves, and the 4:2:2 color space, it really looks like film.

Cinema Vanguard

Caporale made cinema history shooting the very first Varicam feature, *Tattered Angel* (starring former Wonder Woman Linda Carter), in 2001 and then *Sweet William* (with Frank Langella and Laura Allen), which was the first Varicam film to be cut with Apple's Final Cut Pro workflow via Firewire (see Chapter 7).

As a front-line soldier in the digital cinema revolution, Caporale was determined to alert other aspiring filmmakers to what new digital filmmaking tools can do. He began teaching at "VariCamps," which are conducted by an organization known as HD Expo as separate three-day seminars in various major areas around the country, including Dallas, Chicago, Denver, Atlanta, Los Angeles, New York, Washington DC, and in Connecticut.

"I and the other instructors are committed to spreading the faith as far and wide as possible by educating Varicam users on the technical use of the camera," he explains. He also conducted a road show tour of 24p seminars for the better part of a year that included tradeshow events, festivals, and dealer seminars. Caporale wrote color set-up files for both Panasonic's Varicam and its SDX900 camera (see Chapter 6); those files are included on a CD sold with the acclaimed *Goodman's Guide* book for that camera.

"I'm frequently asked why I share this information (set-up files and so on)," Caporale states. "After all, I'm enabling potential competition. But anyone who understands the democratization that the digital cinema revolution represents already knows the answer to that question."

Caporale is also, however, careful to remind filmmakers that all the pieces for a truly decentralized, democratized production future are not yet in place.

"The digital cinema revolution still has to occur in distribution," he cautions. "What's happened is that we've got an upside-down kind of marketplace with far too many sellers and not enough buyers. We have the ability to make movies but we don't have the ability to get them shown. Studios own theaters and distribution, and they control what gets shown. But if we can distribute via satellite to independent theaters, then anything is possible."

Caporale also cautions that *independent* doesn't necessarily mean *better*.

"I've been to a lot of festivals, and I've seen a lot of dreck," he admits. "Of course, you can make dreck at any financial level. You have to have quality actors, you have to have a quality team, and if so, the quality will get noticed. I have every confidence that, as with any other system, quality will rise and we'll be able to enjoy a broader cinema experience.

To that end, Caporale has formed a new company, 24P Digital Cinema, based in Cincinnati, to provide equipment and expertise to do-it-yourself filmmakers as well as full production support, individual on-site training, and consulting. His web site (www.24pdigitalcinema.com) features sample footage shot with the Varicam, SDX900, and DVX100A. Also on the site is a full 720 by 1280 resolution 18-minute short *American Made*.

Caporale continues to work on digital features as well. Recent credits include digital consultant on the indie film *Jimmy and Judy* (starring Edward Furlong), which was photographed with the Panasonic DVX100A.

Although he sells access to the tools of cinema, Caporale emphasizes that clients have to bring their own creativity. "I love filmmaking and have always considered myself as a visual person," he says. "I am thrilled at the democratization that digital filmmaking brings. But just being an independent doesn't mean you'll have a watchable, exciting, engaging film any more than if you have lots of money and stars. There has to be good storytelling.

"Independent film can often feature more innovation, new ideas, and risk-taking than you see in major motion pictures," he concludes. "And that's what I like about it the most."

The Best Thief in the World

The camera/recording system options for shooting a film digitally grow with each passing year. Some are high-performance expensive technologies specifically designed for making motion pictures. These include the high-definition Panavision Genesis system, the Thomson Viper Filmstream, and the Dalsa Origin (see Chapter 5). Other cameras were originally intended for consumers, but offer features and picture quality so good that filmmakers have used them with great success. These include the Canon XL2, which uses the mini-DV video-cassette format. And then there are those cameras in the middle, originally designed for television news, but also well-suited for filmmaking. Such cameras include the MPEG IMX-based MSW-900P. New York-based writer/director Jacob Kornbluth used this camera to shoot his indy film *The Best Thief in the World* (2004), and in this chapter, relates his experiences with this new digital cinema tool.

Taking Risks

For Izzy and Jacob, it's all about taking risks. Eleven-year-old Izzy's risks involved breaking into neighboring apartments to see how far he could push the boundaries of his troubled young life. Thirty-one-year-old Jacob's risks involved breaking the celluloid ceiling to see how far he could push the boundaries of a new digital filmmaking technology to tell Izzy's fictional story. Each risk, in its own way, paid off. Kornbluth, director/screenwriter of *The Best Thief in the World*, reports he "couldn't be happier with the choice to shoot digitally" using Sony's MPEG

IMX-based MSW-900P camcorder. As for what happened to Izzy, it's all in the movie.

"Digital is becoming synonymous with the ability to take risks," Kornbluth elaborates. "And I think it's because the format democratizes the medium—it's so expensive to make movies—that having [moviemaking technology] in the hands of the people is fantastic and something to be celebrated. In my case, *The Best Thief in the World* is a story that was ambitious and risky; taking some chances and shooting it digitally basically allowed us to make it. I don't know if I'm wired in such a way that you could ever say there's *no* way we would *ever* have made the film *without* digital, but you could probably make that argument."

Hand-Held Mobility

Kornbluth's first feature, shot on 35mm film, was the acclaimed 2001 Sundance film *Haiku Tunnel*, released by Sony Picture Classics. He cowrote and codirected that film with his brother Josh. This time around, Kornbluth is the screenwriter and director of *The Best Thief in the World*, which premiered at the 2004 Sundance festival in the Dramatic Film category. The film is produced by Tim Perell and Nicola Usborne and coproduced by Howard Gertler and Scott Koenig. A Process and Jorge Films production, it was one of the first films to be released under Showtime Network's Independent Films group, which finances movies directly for independent theatrical release. This coming-of-age family drama, which stars Mary-Louise Parker and young Michael Silverman (as Izzy), was shot by DP Ben Kutchins in practical locations in New York City. Or, as Kornbluth explains it, "in real apartments, real neighborhoods, and with real thugs outside trying to break into our truck." As such, it presented unique production challenges.

"Ninety-five percent of the film is shot hand-held," Kornbluth relates. "We shot during the summertime in New York city inside apartments with the air conditioning shut off so that it wouldn't mess with sound. By the end of a take, Ben would take the camera off his shoulder and he'd look like we just dumped a bucket of water on his head."

That being said, it was the Sony MSW-900P camera that enabled Kutchins to shoot the intimate, close-quarters scenes that were essential to telling the story of *The Best Thief in the World*, yet also achieve the image quality necessary for theatrical motion-picture release (Figure 12-1).

Figure 12-1

(Left to right) Director of Photography Ben Kutchins with Sony MSW-900P camcorder, Writer/Director Jacob Kornbluth, and Co-Producer Howard Gertler, creators of the digital indy film *The Best Thief in the World.* (*Photo by Thomas Duncan Photography.*)

"A lot of the advantages of shooting in video showed up in different ways," Kornbluth continues. "You could carry the camera around and get into tight places. We were using film lenses with a Pro35 adaptor, so the camera wasn't light or particularly small by the time we were done rigging it up. But the film certainly wasn't something we could have done with a 35mm camera. And the support-package stuff—including the stock itself—made it a bit more mobile too."

The adaptor Kornbluth mentions—officially known as the Pro35 Digital Image Converter—is made by P+S Technik and sold in North America by ZCG. It enables digital filmmakers to attach any Arri PL-mounted prime 35mm film lens to an SD or HD 2/3-in. camera and obtain what the company terms "the three-dimensional quality of a 35mm film camera on videotape." The Pro35Digital projects the 16mm by 22mm image produced by a 35mm lens onto a specially designed, eccentrically moving ground glass. "It gives an apparent grain to footage and a texture that you really don't get when you're using a standard video lens," Kutchins noted in a Sundance Festival newsletter. "It's another element in front of the lens that could cause focus problems. But I feel like the rewards are pretty huge."

MPEG IMX

The other rewarding technology experience for the makers of *The Best Thief in the World* was the use of Sony's MSW-900P camcorder. Designed for newsgathering, the MSW-900P employs the MPEG IMX format, which records compressed standard-definition digital video at 50 megabits per second (Mbps) using the internationally standardized 422Profile at MainLevel format for production and transmission. In addition, MPEG IMX VTRs can play back all of Sony's half-inch formats, including Betacam, Betacam SP, Betacam SX, and Digital Betacam. MPEG IMX has been widely adopted by such major television broadcast networks as NBC, but it has also found favor among filmmakers as a digital alternative to Super-16mm film. The MSW-900P camcorder not only offers the 16:9 aspect ratio filmmakers prefer, but also high sensitivity for low-light levels, four channels of 20-bit audio recording, and—in the MSW-900P PAL version—a frame rate of 25 frames per second (fps), which is a definite advantage for producers intending to record their finished product onto film at its standard rate of 24fps. The camera also features a digital time-lapse feature. Famed DP David Leitner used it to shoot nighttime Manhattan traffic and create a visual metaphor for frenetic urban life in director Scott Saunders' *The Technical Writer* (2003), the first film ever shot with the MPEG IMX format.

"The DP and I looked at a test we did to see what the IMX camera could do and we were ultimately impressed with how it held up under daylight conditions and how it held up under night conditions," Kornbluth states in preparing to shoot *The Best Thief in the World*. "We did our own test with it to see how it handled these various types of environments and I think we had to worry about them less than if we had used DV, for instance. [MPEG IMX] held up to what our experience with film was."

"I approached it initially pretty skeptically," Kornbluth continues. "I needed to understand how it could be an advantage. I wasn't there just because of the general novelty of working in a trailblazing medium. I needed to figure out a way that it would have some creative advantages for me.

So I put it through a rigorous test, mentally, about how it could be used and in what ways I could use it. A bunch of things appealed to me: It was new and offered an opportunity; it was like a blank slate. It was something we could work on to see how it works best for us without having a body of knowledge in the background. It appealed to that sense that people have been approaching digital filmmaking with for a while. It's the sense of: This is the frontier; let's figure out how to make it work and what's good and what are the true advantages of it.

"I think the main thing that first got me was the idea that I was really looking to get a sort of *verite* feel from the story and I was going to be working with kids that I had a pretty good idea were going to be nonactors," Kornbluth adds. "And I really wanted the sense that the camera wasn't going to be the focal point for the actors, that they would feel as natural and be as unaware of the camera as possible. And the IMX format really seemed to offer those possibilities in a way that film certainly couldn't. And looking back on it, having been through it, I think it really did that for us."

Myths and Realities

Having already directed *Haiku Tunnel* in 35mm, Kornbluth's work on *The Best Thief in the World* provides him insights to comment on how digital filmmaking is different.

"The first thing I'd heard about video from other directors was that it makes you work faster, that there's less time between set-ups and you can make a film more quickly," he observes. "I actually did not find that to be the case for me. If you still want something to look good, you still have to light it. And the lights don't get set up any faster.

"What I did find, however, is that once you set up to shoot a scene, within that scene, you don't have to roll and cut in the same ways that traditional filmmaking requires. You don't have to say 'Cut!' because you don't have to save every inch of film. The fact that tapes last an hour really has a fundamental effect on the way actors can act in front of a camera. You can keep rolling and talk to them for two or three minutes and say, 'Why don't you try that?' right in the middle of a sentence. And then suddenly they're off and acting again. It allows for a bit more of an organic style and the chance to do what I call 'continuous takes.' We could do four or five takes in a row without the crew playing with lights. We could let the actors act in a space for extended stretches that might last 10 or 15 minutes. That was really exciting. That really brought something new to a way of working with actors for me. Also, you can see what you shot just by rewinding the tape if you need to, which is amazing. And you're not reloading the camera, buying stock, paying for transfer" (Figure 12-2)

And what of the low-light advantages of digital, an area often praised by filmmakers experimenting with alternatives to celluloid?

"We shot some night stuff and I was particularly happy with how MPEG IMX reacted in low light, dealing with night images," Kornbluth replies. "We didn't consciously make it low light or push that to any extreme, but when we were forced to for various logistical reasons and I was really pleased with how it responded.

"More important to us was the contrast ratio—no bright windows in the background, and how bright the windows were relative to the interiors," he adds. "We wanted to keep the contrast low and make sure that the difference between the brightest spot in the image and the darkest spot in the image wasn't so extreme. I think the lights themselves—a lot of Kino-Flos—were pretty small units in our case. We knew we'd have a lot more ability to handle the image in post with digital, really manipulate it after we're done shooting. And I think it was something the DP really felt strongly about too, that one way to make it look most beautiful and offer us the most image control was to spend a little bit more time keeping the contrast low in shooting so we could boost it in post-production, and that actually was what took us the time.

Figure 12-2

Jacob Kornbluth directs as DP Ben Kutchins shoots apartment building exteriors with the Sony MSW-900P camcorder.

Post

"We used Sony's XPRI nonlinear editor," Kornbluth states, "and the possibilities MPEG IMX offers once you're done shooting were particularly great for me. The XPRI is designed specifically to work with the IMX format, so you can see what you shot in 25 frames." He adds that being able to see their images without further compression on the XPRI was a special bonus.

"We were basically editing the same image that we shot," he observes. "And also things that are opticals in the film world—like simple attempts at dissolves and different transitions, or slowing things down by overcranking in production—you can do easily in digital post. These things are very liberating; they allow you to make a connection between all of the possibilities you can think of and what you can do. And if you're rigorous about it, digital lets you really push things as far as you can think, which is great."

"Not having to do a separate online before we did our color-correct was really nice," adds coproducer Howard Gertler. "We had to do some TV inserts, and we were able to do them directly in the XPRI, which was an advantage. Honestly, this is the fastest I've ever had a film go between picture lock and finishing for its premiere exhibition. And I don't think you can do that in film, where you're cutting negative and everything just takes longer in making your prints. With digital, the facilities you're working with are flexible and go much more quickly."

"I think it looks great," Kornbluth says of *The Best Thief in the World*'s final image. "We spent a lot of time thinking about the aesthetic and the feeling of the whole movie. The way that the visual aesthetic reflects what the narrative intentions are is something that I'm really pleased with. I couldn't be happier with the choice to shoot with MPEG IMX. Digital's definitely come to the point where if you use it, you can get a lot out of it. I'm really pleased with that."

Interview with Michael Ballhaus, ASC

Budget-minded independent filmmakers were among the first to embrace video camcorders and digital high-definition tape as an alternative to 16mm and 35mm film. Attracted by such factors as affordability, portability (great for *cinema verite*), and compatibility with PC-based editing systems, the tools of digital cinema had instant appeal to "indy" filmmakers.

Mainstream, big-budget moviemakers, meanwhile, are varied in their response to new digital filmmaking tools. While digital intermediates for motion-picture postproduction have become the norm in Hollywood, digital cinematography has seen slower acceptance. Other than a handful of such major directors as George Lucas, Robert Rodriguez, and James Cameron, theatrical movies are still principally captured on 35mm film. Will this change?

The answer lies in large part in convincing the cinematographers—those craftsmen and women whose job it is to actually capture the images their directors envision—that digital cameras and recording formats are worthy alternatives to film. One such cinematographer with experience in both film and digital is Michael Ballhaus, ASC ("ASC" indicates membership in the American Society of Cinematographers, the ultimate certification for all "masters of light"). The author of this book interviewed Ballhaus in May of 2004 to seek his perspective on digital's present—and future—acceptance among cinematographers.

Whether it's an intimate character study, such as director R.W. Fassbinder's *The Marriage of Maria Braun*, or a big-screen spectacle like Barry Sonnenfeld's *Wild Wild West*, Michael Ballhaus, ASC has consistently delivered the creative visions sought by a wide range of directors and scripts. A master of his craft, Ballhaus is as much at home in his native Germany teaching cinematography as he is in Hollywood shooting pictures for Martin Scorsese, Francis Ford Coppola, Robert Redford, Mike Nichols, and fellow countryman Wolfgang Petersen. With scores of films and multiple Oscar nominations to his credit, Ballhaus is one of today's most accomplished and influential filmmakers, having lensed such hits as *Something's Gotta Give*, *Gangs of New York*, and *Air Force One*.

And as if this wasn't enough, Ballhaus also teaches cinematography at universities in Berlin and Hamburg. "I have to know what's in the future for my students," he explains. "I'm always interested in trying new things, experiencing new technologies, and knowing how fast they're developing."

Putting this belief into practice, Ballhaus shot a digital test film, *Triell*, in early 2004, using the Thomson Viper FilmStream camera, which records motion imagery at what is known as "4:4:4 RGB 10-bit log." This data-rich format is said to rival the detail, nuance, and subtlety that's possible with film imaging. The data were recorded on the Director's Friend, a computer disk-based image-capture system. A company called Band Pro Film and Video/Munich provided technical support on the project devising a workflow to handle the Viper's 4:4:4 FilmStream data in the offline edit and ensuring that Ballhaus' footage had synchronized time code for online editing. *Triell*'s final images were output to 35mm via an ARRI Laser directly from the corrected raw data in 4:4:4 as well as Sony HDCAM SR and HDCAM cassettes.

"I'm pretty sure that it will look pretty damn good," Ballhaus anticipated prior to seeing the final results. "If you are used to film cameras all your life, it's just a change, but I am open to these new developments, and as long as these cameras are good and getting better, I'm open to work with them."

The following questions were posed to Ballhaus by the author of this book, with additional questions provided by Hollywood director of photography James Mathers and Bob Zahn, President of Broadcast Video Rentals, in New York.

How would you describe the work of a cinematographer?

That's a complicated question. A cinematographer is the one person who puts the vision of the director on the screen, basically. And whatever he can add to that with his technical knowledge and his own fantasy that helps to make the film as visual as possible, that's the job of the cinematographer.

Figure 13-1

Michael Ballhaus, ASC with Thomson Viper camera. (*Photo by Regina Recht.*)

Is it accurate to say that cinematography is 'painting with light?'

Yes, it is painting with light, but that's not all, because the light is not everything. Light is the same for a photographer. A cinematographer has to know more than just painting with light. He has to think about the movement, he has to think about what comes together when he shoots a sequence, that he knows which frames will meet. What's the rhythm of a scene, and how can he tell the story in the most visual way, the most dramatic way, to photograph a scene. And that is much more than painting with light.

Does cinematography only involve the use of film, or are digital cameras also valid?

Oh, they are absolutely valid for cinematography, and more and more so because technology develops really fast. I had the chance to shoot a short film [*Triell*] in Munich using the Thomson Viper FilmStream camera and it was an additional experience because the quality is getting really good and it comes pretty close to film.

You are impressed with the Viper FilmStream camera?

Yes I am. I think it comes a little closer to what a cinematographer needs.

In what ways?

In the ways of how to handle the camera, the ergonomics of the camera body, how it's formed, and all the things that a regular film camera needs to work. And I think they built Viper a little closer to what a film camera looks like and works like.

What about the quality of the images you made with Viper?

I was impressed with the quality that, at the end, you see on the screen. There's still some things that are a lot easier—a lot more common and better so far—with film, because film cameras are very durable. You can shoot in the rain, you can shoot when it's cold, when it's hot; they always work. And you can work under the most complicated circumstances.

I think these digital cameras are more sensitive. And so far, it's harder to work with them. Also, they are not as movable because they have this thick cable attached to them. So, there are still things that are better with the normal film camera, but I think the quality of the digital images is very good.

Would you say that those digital images are on a par with film images?

Not quite, I would say, but it's coming pretty close.

You're known for capturing very challenging cinema imagery; can you give an example of a shot you did using film that would have been facilitated more easily today with digital technology?

Easier, yes, but not better. When I shot *Dracula* with Francis Ford Coppola, there was a scene where Keanu Reeves is sitting at a desk writing a letter, and Dracula (played by Gary Oldman) is behind him, leaning over his shoulder. And you see Dracula's shadow on the back wall. And this shadow moves as he moves, but at one point, the shadow has its own life and tries to strangle Keanu Reeves. It puts its hands around his neck and strangles him. But it's just a shadow.

Today, you would do that digitally because it's a lot easier, although more expensive. But what we did in 1992 was use a second Dracula actor behind a screen and project his shadow off the first Dracula. The second Dracula actor, who wore an identical Dracula costume, had a monitor to watch every move that the first Dracula (Gary Oldman) did. And he matched each move exactly until the point when his shadow appeared to take on its own life. The effect was challenging in a way, but it was easy and adventurous, and it looked great. It was also not expensive.

We did a lot of in-camera special effects on *Dracula*. In my first meeting with Coppola he said, "Our role model for this movie is the 1922 *Nosferatu* by F.W. Murnau." So we tried to do as much in the camera as possible. We did things like double exposure and dissolves in the camera, and we ran the film backwards for special shots.

What inspired you to want to do the test film Triell?

I'm always a front-runner. I was one of the first cinematographers in Germany to introduce digital cinematography to students, and I'm always very interested in where the technology is going.

What about monitoring the digital image during cinematography with the Viper FilmStream, the actual viewing of the colors?

We had to filter the images because on the monitor, they looked green. So we had to use a filter to take the green out. Then you could really judge what you had captured.

What about the Director's Friend recording system; did it make things easier or more difficult—as opposed to using film?

In a way it's easier because you can play it back. You don't have to look [at] dailies, you know what you have in the can, so to speak. And that makes it safer.

But on the other hand, we had to reshoot half a day because we had some electronic problems in the digital recording. So I would say it's more sensitive than the film and the film camera. But digital cameras are only at their beginning; they will develop fast and get better every time. I'm pretty convinced that in a couple of years, they will be as flexible and as good as film cameras.

Consumer cameras are getting much better, and you now see motion pictures that are shot on consumer cameras that are excellent and good quality. And when that works for the story, it's great to use those small cameras. Panasonic has a new camera, the DVX 100, with a 25p mode, which—transferred to film—looks pretty good, I must say.

Cameras such as the Canon XL2 are democratizing filmmaking. Is that good or bad?

For me it's a good thing, because it opens the market for new talent. I recently saw an extraordinary film at the Max Ophuls Film Festival, in Saarbrucken, a festival for first-time directors. It was called *Mucksmäuschenstill* and it won every possible prize from the jury, the audience, the actors, the students, and the TV station. And it was done for $60,000 with a consumer camera. It's just a great way to make movies if you have a brilliant idea that works for that system. You cannot shoot *Gangs of New York* with such a camera, that's for sure. But there are many other stories that you can do with these consumer cameras that are excellent, and that opens a wide range of talents to our craft.

Is that a bad thing for professional cinematographers, to have this new competition?

No, it's not. I like competition. It means we all have to be better at what we do.

What do you say to cinematographers who fear new digital technologies?

It's hard to say. If you work all your life with film cameras—and film is something wonderful, as we know—and you're raised in this tradition, then I can understand. But when you're teaching, when you work with young people—students—who want to become cinematographers and directors, you have to be open to these new technologies that possibly are their future.

Photorealistic computer-graphic imagery is another major means of digital 'filmmaking' these days. Are there special challenges for cinematographers who have to shoot 35mm film for later compositing with computer-graphic imagery?

Yes, there are special challenges, but, I must say, it's a little boring because you are not creating the images anymore. You have a storyboard and you have an animatic that they do of these scenes where you just shoot the foreground and then the background is put in later. In your animatic, they tell you where to put the camera, what lens you have to use, the distance to the object—all these things—and then you have to light it in a way that it's easy for them to do the composite later. It's very technical for a creative cinematographer who wants to be in charge of everything. It's not inspiring. You're not a master of the imagery anymore; a lot of it is done later in the computer. There are other DPs who might like that, but for me personally I'm not so crazy about it.

Does it turn you from being an artist into being more of a technician?

Yes, exactly, that's exactly it. And I'm not a technician. I feel I am an artist. But then the end result is fine, it's okay.

What would you say cinematographers must do to ensure that new digital technologies don't adversely affect their craft?

Manufacturers need to work closely with cinematographers. Sometimes they may put too much technology into the camera, things that you would never use; so you discuss it with them, and they should listen to you.

I was the first guy who saw the ARRICAM, which was about a year before it came out. I looked at it, and we discussed what I would like to have changed, what I thought is and is not necessary. The guys at ARRI really do listen. The ARRICAM is developed by Fritz Gabriel Bauer, who was a cinematographer himself, and he's just brilliant. He does really know what a cinematographer needs and wants. So they are pretty damn good, these guys.

The new ARRICAM is fantastic, I worked with it on my last movie and it's the best camera I ever worked with. It's just amazing. And Arriflex is developing a digital camera called the D20, which is, right now I think the most interesting development on that level of digital camera. I haven't had the chance to work with it, but I will pretty soon check it out. It has the best viewing system because it's an optical viewing system. The D20 is based on the ARRI 435, which is a brilliant camera. The imaging chip it uses is the same size as a 35mm film frame, and you can use the regular lenses...with the same depth of field that you have on film. These are all things that make it look much closer to film than the digital cameras with smaller chips. And so I'm very interested in this development.

Do you think film will ever go away?

Digital images and film images will exist next to each other for a long time, I think, because there are still millions of projectors all over the world that still need to be fed with film. When I first saw the first digital camera, which was about seven years ago, I thought then in two years it would be over with film. But we're still working with film and loving it. I hope it doesn't go away too soon, because I like film a lot. But I think in the future, when there are digital projectors in every movie theater, then—slowly—film may come to an end. But I have been wrong before. As I said, however, I'm open to the new developments.

Digital Projection—Films Without Film

Despite all the progress in cinema technology made during the past 100 years—despite advances in cameras, lenses, film chemistry, widescreen movies, multichannel digital sound, computer animation, and so many other innovations—the basics of movie projection really haven't changed at all since the days of Thomas Edison. Film is a long, flexible transparent ribbon on which are printed hundreds of thousands of still images. Unspooled from large metal reels, film is mechanically transported via its sprocket holes so that each film frame passes momentarily before an intense light source, which projects those images onto a screen. A relatively simple technology, film projection works extremely well, which is one reason why it's been in use for so many years.

The amount of picture information that can reside within a 35mm film frame is equivalent to at least 2,000 lines of image resolution (today's HDTV offers only 1,080 lines). Film images are robust enough to be projected and displayed onto theater screens at hundreds of times their original 35mm-wide size and still provide a high-quality, larger-than-life viewing experience. Projected film offers a rich palette for moving-image communication and a potential bonanza for creative filmmakers. In addition, a century of 35mm film exhibition around the world has made the format the only truly global imaging standard, one that has outlived scores of videotape formats used in television.

There are probably at least 100,000 film projectors in theaters worldwide. They work. So why change anything? Pretty much for the same reasons that vinyl long-playing phonograph records were replaced by digital compact discs.

As anyone who has seen a major motion picture on the first day of its release can testify, film projection can produce beautiful images, assuming the projector itself is set up properly. Unfortunately, not all ticket buyers make it to the grand opening, or even the day after. And every time a film print passes through the projector, it fades slightly. It picks up scratches. Dirt accumulates on the print. The sprockets widen from stress, causing the image to weave. And by the time the film nears the end of its run, its project-ed image quality will be a far cry from its first few screenings. As theater goers, we've lived with this inevitable deterioration simply because there was no alternative—until now. That alternative is digital cinema projection.

There are many reasons to convert to a digital-cinema delivery sys-tem, one that delivers films to theaters via satellite or hard-disk courier instead of reels of film. The most compelling reason, however, is to use a filmless projector. The advantages of such a system are obvious: no wear and tear on the film, no jitter or wobble, no problems with dirt and scratch-es, and no loss of picture quality from fading. Such problems can be elim-inated forever with the use of an all-electronic imaging system. In fact, it may well be that digital projection is the best way to preserve the true beauty of 35mm cinematography, given that film prints degrade fairly soon. But although doing away with projected film may sound like a desir-able and obvious choice, the truth is that projection systems for digital cin-ema have a few hurdles of their own to overcome.

Setting the Standard

For all of its physical limitations, motion-picture film does have one attribute most electronic imaging systems are hard-pressed to equal: a wide visual dynamic range. The most important attribute of any image is its range of tonal shades (or luminance values) from the deepest black to full white, along with all of the shades in between. The more shades of gray (or steps) we can see in an image, the more detail it offers us. And it follows that each step of gray represents another shade of color that can be represented.

In this capacity, motion-picture film remains the "king of the hill." Its ability to capture a wide range of luminance values in scenes with high contrast is combined with a wide range of colors (or gamut) to present bright, contrasty images with a high degree of photorealism.

This is a tall order for any digital cinema projection system to fulfill, but it's clear that some manufacturers are getting closer with each passing year. Indeed, some in the industry maintain that the best digital cinema systems are already "good enough" and have exceeded the quality of motion-picture projection in the average multiplex theater.

If we were to photograph a smooth grayscale—a "ramp" from black to white—we might easily come up with over 1,000 values of gray. Add in three primary colors of red, green, and blue, and we now have a palette with well over one billion possible shades of color. Contrast that with the typical computer display card, which might only be capable of generating just under 17 million shades of color (256 shades of red x 256 shades of green x 256 shades of blue = 16.7 million colors). Although that number might seem like a lot, it clearly isn't enough to capture the tonal range present in motion-picture film.

Our ideal digital cinema imager, then, must be able to reproduce as many shades of gray as possible, and do it while not compressing any part of the projected grayscale. That means that "black" is black, not a low gray. It also means that we should be able to clearly see the nuances between several shades of "near" white, not a washed-out bright area without detail.

Just Add Color

The problem gets more complicated when we add color to our ideal digital cinema grayscale. Motion-picture film is typically projected using light from a "short-arc" xenon lamp. Xenon is a rare gas that, when stimulated, gives off spectral energy that is almost equal in intensity from longer-wave red, orange, and yellow light to shorter-wave green and blue light.

Assuming the film lab has done its part and has processed the film to yield the colors that a director and cinematographer originally intended, the result is a representation of color that approximates what we see around us every day. Flesh tones appear normal, pastel shades are rendered with subtlety, and colors are neither over- nor undersaturated—unless deliberately manipulated by the director and cinematographer for effect.

It stands to reason that our digital cinema system should also use an equal-energy lamp, and in fact, all the systems currently offered for digital cinema projection do employ such lamps. But there's another hurdle to clear.

The digital cinema imager itself must have absolute color purity. That is, when the imager projects a black and white image, it should have no evidence of any color tint or shading. If a color image is projected, it should remain consistent across the entire screen area.

Don't Skimp on Details

The third attribute of our ideal digital cinema imager is resolution—and lots of it. The ability of motion-picture film is mostly limited by its grain structure and film speed (exposure sensitivity). Films with moderate to low sensitivity have finer grain structures and can capture a high degree of image detail. The ability of motion picture film to resolve fine detail can be measured in "line pairs."

In a digital cinema projector, the resolution of the projected images is limited by the physical array of pixels in the imaging devices. This physical pixel resolution (also referred to as "native" resolution) can easily be measured, and is usually expressed as horizontal pixels by vertical pixels, not line pairs.

A typical motion-picture film stock might have the ability to capture well over 4,000 line pairs of information. In contrast, the newest digital cinema projector might be equipped with devices capable of displaying only 1,920 by 1,080 pixels. In an ideal world, we'd like our digital cinema projector not to be the weak link in the chain when it comes to resolution!

Of course, there are a finite number of line pairs of pixels that can be perceived by the human eye. Your ability to resolve picture detail and tell a low-resolution image from a high-resolution image has to do with how far you sit from the screen, how large the screen is, how sharp the image is, and how well the projector handles grayscales and color.

As it turns out, meeting the grayscale, color, and sharpness benchmarks is easier said than done. That's why many systems proposed for digital cinema projection need much more work before they are ready for their moment in the spotlight. Let's now take a look at the imaging technologies that are considered strong contenders.

Liquid Crystal Displays (LCDs)

LCD technology comes in many flavors. The underlying principle of liquid-crystal displays is their ability to act as light shutters, just like window blinds. By applying different voltages to a pixel containing liquid-crystal molecules, the degree to which polarized light is blocked or passed is changed.

In effect, the LC pixel becomes a grayscale "light modulator." If we were to view the pixel with a magnifying glass, we would see it changing from light to dark and in-between. This principle of liquid crystals was first observed in the late 19th century, but was not commercially harnessed until the 1970s for use in desktop calculators and wristwatches.

The type of liquid-crystal display we are most familiar with functions as a *transmissive light modulator*. That is, light from a constant source (such as a fluorescent lamp) passes through the LCD to our eyes. Tiny color filters applied to the pixels create the red, green, and blue color mixing necessary to complement the different luminance steps.

Transmissive LCD panels also come in tiny sizes suitable for everything from ultraportable business projectors to large, "light-up-the-room" models for concerts and special events. All, however, suffer from the same limitation: an inability to display a deep, rich "black."

Just as a window blind can't completely cut off all light, these LCD panels also allow some light to pass through in their "off" state. While significant advances have been made in solving this problem, digital cinema interests have largely dismissed this technology as unsuitable for their needs. The fact that LCD panels have a hard time maintaining a high level of color purity across their surface only adds to the problem.

Transmissive LCD panels are also limited in terms of pixel density. A typical LCD panel with the 1,920 x 1,080-pixel array mentioned earlier measures about 1.8 inch diagonally, much larger than an equivalent 35mm film frame. That would mean that special optics are required and the projector housing is increased considerably in size.

Liquid Crystal on Silicon

Liquid crystal on silicon (LCoS) shares many of the attributes of transmissive LCD panels, but it works in an entirely different manner. Light is reflected through the LCoS panel to modulate it from black to white. That's necessary because the LCoS panel is opaque on its backside, where the controlling electronics for each pixel are mounted.

LCoS panels also use different types of liquid crystal molecules to shutter light, and can achieve much higher pixel density for a given panel. A 1,920 x 1,080-pixel digital cinema projector would require LCoS panels measuring no more than a .8 inch diagonal—less than half the size of transmissive LCD technology.

The optical path of an LCoS projection engine can be somewhat complex, but basically involves three monochrome LCoS panels precisely mounted to a special prism known as a *polarized beam splitter* (PBS). This device passes light in two directions at once, a trick made possible by first polarizing it.

Each panel has special color filters in front of it that respond only to red, green, or blue light refracted by the PBS. The LCoS panel modulates a black and white image that is colored by the nearby filter as light reflects back into the PBS. Identical copies of the image are also modulated by the remaining two LCoS panels and colored by the PBS filters.

When the resulting red-and-black, green-and-black, and blue-and-black images are combined in exact registration, the result is a full-color image that then passes through the projection lens and onto the screen.

As you might imagine, the issues that apply to transmissive LCD technology also rear their heads here. Black levels are a problem for LCoS projectors, although not as much as with transmissive panels. Color purity and uniformity must also be monitored, as the liquid crystal pixels can create a noticeable shift from red to blue across an image. Image contrast is also an area that needs more improvement.

Still, LCoS shows great promise because it is capable of much higher resolutions than are practical with transmissive LCD. Companies such as JVC and Sony have shown LCoS projection systems with 1,920 x 1,080 native resolution, and have also demonstrated systems with over 4,000 horizontal pixels.

JVC's technology goes by the name Direct Drive Image Light Amplifier (D-ILA, for short), while Sony has adopted Silicon Xtal Reflective Device (SXRD). Both are simply different designs of LCoS technology. Other companies actively engaged in LCoS manufacturing include Hitachi and Intel.

From Analog to Digital

For all of the benefits of LCoS technology, it is still very much an analog imaging system susceptible to heat, noise, and other factors. Another way to design a digital cinema modulator is to use a 100 percent digital system, such as Texas Instruments has done with its Digital Light Processing (DLP) system. At the heart of DLP imaging is the Digital Micromirror Device (DMD), a specialized reflective semiconductor with thousands of super-small mirrors etched into its surface. These tiny mirrors tilt back and

Figure 14-1

High-quality digital cinema projection has become essential for high-end postproduction and digital intermediate facilities, such as Randall Dark's HD Vision Studios, in Los Angeles.

forth over a total of 12 degrees in response to control signals from their underlying semiconductor drivers (Figure 14-1).

At first glance, a system with mirrors that can only tilt on or off doesn't seem much more useful than a light switch! The trick is to use humankind's persistence of vision and the super-fast on-off cycles of electronics to fool our eyes into seeing continuous grayscale images. That's done with a process called *pulse-width modulation*, which sounds more complicated than it really is.

Imagine you can turn a light switch on and off as fast as you need to. At normal speeds, our eyes simply see the room going dark, then bright, and then dark again. But if we speed up this on-off-on cycle sufficiently, the room may appear not to be dark or bright, but somewhere in between. By precisely setting a number of "on" cycles and "off" cycles in a given time interval, our eyes can now see many different levels of brightness in the room—ergo, we've created grayscale images out of nothing more than simple but very rapid operation of a light switch (assuming we could actually turn a wall switch on and off that fast).

That's all there is to DLP. Like an LCoS panel, a DMD is a monochrome light modulator. Three such DMDs must also be mounted to a polarized beam splitter with color filters to create the required, precisely aligned red, green, and blue component for a full-color image. Unlike an

LCoS panel, a DMD is 100 percent digital in its operation and largely ignores the effects of heat and noise. This means that it can create a very pure shade of any color without a color shift across the screen. The black levels of DMDs are also much lower than that of LCoS or LCD technology, which lets us expand our grayscale and subsequent palette of colors.

At present, TI's DLP technology (branded under the name "DLP Cinema" and also known as the "black chip") is a strong contender for digital cinema projection systems along with JVC's D-ILA and Sony's SXRD. More than 100 DLP Cinema projectors have been permanently installed in major theaters and multiplexes worldwide and are used to show first-run movies each and every day. Currently Barco, Christie Digital, and Digital Projection Inc. (DPI) are the three companies that license TI's DLP Cinema technology in the digital cinema projectors they manufacture.

TI's highest-resolution DMD is currently offered with 2,048 x 1,080 native resolution and no higher-resolution demonstrations have taken place as of this writing. There is, however, nothing in the structure of the DMD that would limit its ability to be manufactured with higher pixel densities comparable to those achieved by JVC and Sony.

Grating Light Valve

Many other approaches have been offered to digital cinema projection. Some of these "can't-miss" systems have come and gone quietly, but others continue to intrigue. One of those is the grating light valve (GLV), a technology first developed by Silicon Light Machines/Cypress Semiconductors.

The GLV device looks like no other imaging semiconductor. Thin, flexible ribbons are suspended over a substrate of silicon transistors. These ribbons actually flex in response to the signals from those transistors, and in doing so, allow light from a coherent light source (a laser) to pass through and onto the screen.

The entire device resembles a tabletop musical finger harp, and it is immediately apparent that the only physical resolution that can be measured is in the vertical plane (i.e., the number of ribbons). The GLV's horizontal resolution is a function of where light from the laser passes through to hit the screen. In effect, a 2K GLV would have a physical resolution of 1 x 1,080!

Sony has made an agreement with Cypress to develop this technology into a workable digital cinema projection platform. Three different lasers and corresponding GLV devices are required to make it work (for red, green, and blue image channels) and the lasers must be recombined in precise registration on the screen, a job made somewhat easier by the fact that they are already tightly focused.

The tricky part is coming up with a blue laser bright enough to make the system practical, and also reducing the size and weight of the associated power supplies for each laser. For that reason, GLV is not expected to be a practical digital cinema projection platform for several years.

Ready for the Audience

Much debating and studying continues around digital cinema projection systems. It seems that no matter how much is done to improve them—regardless of the imaging technology used—there are always groups who persist in saying that digital image projection "just isn't good enough yet to show the full beauty of 35mm film."

The truth is, however, that digital projection is quite "good enough" for many parts of the world. In particular, cinemas in Asia are quite enthusiastic about digital cinema projectors and are installing them in a variety of venues. In fact, in some locations, lower-cost projectors intended for the rental and staging market are being put into digital cinema service.

An argument can be made that the limitations of film projection detailed at the start of this chapter are such that the adoption of digital cinema projectors using the current state of technology is a sensible and practical thing, never mind the cautions and protestations of numerous industry standards committees and vested interests in Hollywood.

As usual, however, the marketplace—and not committees—will determine when it is prudent to move away from projected film and into the realm of digital cinema projection. The average moviegoer really can't tell the difference today between the high-end digital projection systems offered for sale and a casually maintained film projector (as so many of them are these days). In fact, that same moviegoer will likely prefer the stability and cleanliness of the digital version over a spliced, dirty, and unstable film print that has been running several times a day for a few weeks. And that is the main benefit of digital cinema projection—to ensure that the last screening has all the picture quality of the first screening, an important consideration given the ever-increasing price of admission.

Texas Instruments' DLP Cinema

"An amazing thing happened in late November 1998," recalls *Star Wars III: Revenge of the Sith* producer Rick McCallum. "I got a call from Doug Darrow at Texas Instruments. He asked us if we would like to see a first-generation demonstration of their new digital projector. We quickly set up a demo at Skywalker Ranch. When we saw the results, we went absolutely nuts. George [Lucas] was so enthusiastic that he wanted us to transfer *Episode I* to show in theaters on a trial basis as soon as the film would be released. This was the final link in the evolutionary chain. The results from the four theaters where we digitally projected the movie were outstanding. Audiences loved the sharpness and detail."

Much has happened since McCallum made his remarks, at a 2002 ShoWest motion-picture exhibitors conference. He and director George Lucas shot two more *Star Wars* films, but used Sony's 24p CineAlta digital HDCAM tape format instead of 35mm film. Other leading film directors have also embraced digital production, and the number of digital movie theaters has continued to grow worldwide, with studios such as Disney, Fox, Miramax, Universal, and Warner Bros. supplying digitally mastered versions of their films. There are now over 100 digital movie theaters around the world, and Texas Instruments (TI) continues to be at the forefront of thrilling audiences with the dirtless, scratchless, and crystal-clear clarity of digital projection.

The secret to TI's digital projection technology is known as the DLP Cinema "black chip," which capitalizes on the company's decades as a leader in semiconductor design and manufacturing. "DLP" stands for Digital Light Processing; at its heart is the Digital Micromirror Device (DMD). Measuring slightly more than one inch square (within a few millimeters of a 35mm film frame), the DMD contains 1.3 million microscopic aluminum mirrors individually hinged on a black background to tilt at 5,000 times per second. Three DMDs (red, green, and blue) illuminated by a projection lamp create the 1280 by 1024-pixel resolution images that are revolutionizing the motion-picture viewing experience. DLP Cinema technology is licensed to only two companies, Barco and Christie, which manufacture theatrical projection systems.

Doug Darrow is TI's Business Manager for DLP Cinema Products. A 19-year veteran of the company, he spends much of his time in Hollywood, where his achievements have included working with Disney on the creation of a field demonstration project that launched prototype digital cinema systems into most major exhibition markets. The following interview was conducted in May of 2002 with the author of this book. Since that time, TI has introduced an improved DLP Cinema DMD, which provides 2,000-line ("2K") resolution digital cinema projection in an increasing number of movie theaters worldwide.

Texas Instruments is primarily known as a maker of semiconductors—DSPs, analog chips, calculators; the DLP Cinema chip almost seems like a departure from the company's core business.

It's not. TI has been in the semiconductor business for many years, and is in many ways responsible for it. Jack Kilby [TI's former Assistant Vice President and the winner of the Nobel Prize in Physics in 2000] invented the integrated circuit nearly 50 years ago, and TI has long been an innovator in semiconductors and semiconductor processing. The Digital Micromirror Device was originally invented in 1977, and it spent a long time in the R&D labs, being considered for a variety of applications that ranged from the military to printing.

In the early 1990s, DMD technology had reached a point where it could be "productionized," taken out of the R&D lab, and moved into something that could have a business wrapped around it. And as TI started developing the technology and putting it into production, the decision was made to focus on display applications, where the real opportunity for growing a business resided.

We started doing "road show" demonstrations of DLP during the mid '90s, as we sought to interest end-equipment manufacturers into adopting the technology. At that time, nobody had been using in any serious way a reflective light-modulation solution like DMD, which uses an array of tiny aluminum mirrors.

Once DLP technology went into production, we applied it to the high-brightness category of video projection. The projector industry saw DLP as an emerging "best-in-class" technology, and we said why not tackle digital cinema? Digital cinema was always thought to be the ultimate and most difficult application for video projection.

Cinematographers have very exacting demands in terms of contrast ratio and the quality of projected film. How has TI addressed their concerns?

I'm going to answer that as more than just a question about contrast ratio. When we first kicked off our effort five years ago for digital cinema, we realized very early on that we did not have the understanding to make the-atrical-quality images. And the only way we were going to get that under-standing was to go to the experts (Figure 15-1).

Figure 15-1

Doug Darrow, Texas Instruments' Business Manager for DLP Cinema Products, with characters from *Star Wars* films during Hollywood premiere of *Star Wars Episode II: Attack of the Clones*. Stormtrooper at left is holding a Texas Instruments DLP Cinema DMD chip. (*Photo © Ryan Miller/Capture Imaging.*)

Our first stop was the ASC [American Society of Cinematographers], in Hollywood, and we went there with a lab prototype built on an optical bench that had to be completely disassembled and rebuilt in a projection booth on the Paramount lot. We invited in cinematographers, and we started working with Garrett Smith over at the Postproduction Group at Paramount, who's an associate member of the ASC. We asked the cinematographers what they thought of the pictures. We knew this technology wasn't ready yet. We asked, 'What are the things we need to do to make it better?'

And they started giving us feedback. Every six months or so, we'd return to L.A. with a new prototype. More people had heard about what we were doing and wanted to come see it, give us input, feedback, point us in the right direction, help us prioritize our development efforts—and that's how we really got the key lessons in creating a projection solution for digital cinema. This went on for years. In fact, our first production product came out in November of last year; we've been in development basically for almost five years.

What role did the studios play?

When we had something that was close to the right kind of a solution, we moved to the studios to get them involved in the process of helping collaborate on our development efforts. We involved film-distribution people, both on the technical as well as the business side. At that time, Disney was very proactive;, they had actually been working on their own technology for digital cinema. Several other studios—Warner Bros., Sony Pictures, Miramax, and Paramount—quickly followed; they wanted to collaborate with us on the development and discussion of what was required in the digital cinema projector.

The value proposition varies depending on the group you're talking to, but to the creative—to the director—the central value proposition is that we deliver an experience that's the one they intended. Whatever they were going for in the look of the movie is faithfully represented not only in every screen, but also in every showing.

How did the studios view digital cinema early on?

The studios viewed digital cinema as an exciting opportunity for their business, not only from the standpoint of potential cost-savings, but also as a strategic imperative for theatrical exhibition.

Theatrical exhibition has had only a handful of innovations over its 100-year reign. Even though new and different forms of entertainment media are being delivered to consumers in a variety of ways today, the theatrical exhibition experience has changed very little. Studios that are more progressive in their thinking want to create higher value, unique experiences in theatrical exhibition. They want to improve the quality and consistency and bring more patrons to the theater. Those new cinema experiences can also take a variety of forms we can speculate on. Digital cinema will spawn a whole series of innovations that may range from linear one-way experiences like we have today to interactive and collaborative experiences. Today's movie theaters, which essentially show new-release movies on Friday and Saturday nights, could become exhibition complexes used in new and different ways in the future.

Getting back to contrast ratio, what about concerns that digital projection still doesn't render blacks accurately enough?

The DLP Cinema technology with our DMD chip is a very efficient light modulator and the things that degrade black level primarily are scattered light. And so as we look at the areas on the chip that scatter light, we can improve the contrast ratio rather dramatically. We have plans to continue to make contrast-ratio improvements, or in this case, reduce the black level as much as possible. That's central to our development and a very high priority because we view it as the most important thing to improving image quality for many applications—but most importantly for digital cinema. So that is an area we focus a lot of our attention on.

Today we have a solution that is very comparable to the presentation of movies in movie theaters from the standpoint of contrast ratio. Nevertheless, to be able to represent a 'look' that all cinematographers are going to completely endorse, we still have to make improvements in contrast ratio to satisfy the 'golden eyes' of Hollywood. Of course, these eyes typically see first-generation prints from the negative, not the fourth-generation high-speed contact prints everyone else sees at their local multiplexes.

The first DLP Cinema projectors were manufactured by TI itself. Why?

It was to seed the market. Barco and Christie are strategic DLP Cinema partners today, but when we originally got started, we didn't have any partners. We put two systems into theaters, one on the East Coast, one on the West for *Star Wars I* in 1999. Hughes/JVC also set up a system on each coast, so there was a total of four digital cinema systems running at the same time during that summer.

Soon after that, Disney asked us for more systems for *Toy Story 2*, and we launched six more in theaters in November of '99 and then six more in December for *Bicentennial Man*. Then we started an effort with Technicolor, Disney, and several exhibitors to deploy more prototype TI projectors into movie theaters, and that effort scaled to about 32 and included Europe, Japan, South Korea, the United States, and Canada. That was really the first proof of concept, the goal being to put this technology in a real-world application and see what real movie goers think of digital cinema—digitally projected movies.

How were those installations paid for?

Disney paid part, the exhibitors paid part, Technicolor paid part, and TI did. It was really a collaborative effort to look at deployment on a fairly small scale, to prove the reliability and performance of the technology, and to start to think about an infrastructure that could handle releasing digital movies.

Were all of them fed by QuVIS QuBit servers?

Yes. We worked with QuVIS early on because we saw they had technology that could be applied to this space. They had an HD video server with a wavelet compression engine that wasn't as expensive as some of the higher-end solutions. It served the digital cinema space fairly well. So we spent a lot of time with them as they got their solution up to speed for the first wave of digital cinema. The track record of the entire system (playback and projection) was remarkably successful. If you add up every single problem—technical things, operator errors, anything—failures were less than one percent for any reason in all of the showings that were run.

Now Barco and Christie are making DLP Cinema projectors, many more digital screens are going in in time for *Star Wars II*, there's additional companies building robust servers, and several more studios are releasing movies digitally.

How did the Barco and Christie relationship come about?

Barco is a very strong engineering company that has been in video projection for decades. They're used to building high-performance solutions, not just projection systems. So they were really a very logical choice. They were very aggressive with DLP, and so it made a lot of sense for them to be involved.

Christie is the largest film projection company in the world, and they know the market really well. They acquired a company called Electrohome, which was one of the first partners with TI in building high-end three-chip DLP projectors, so they were a natural fit as well.

We originally signed up three DLP Cinema partners, which we felt was the right number. The third was Digital Projection, which was later acquired by IMAX. That company had some struggles, as did most exhibition companies over the last two years. Digital Projection went through a management buyout, and through that process, we terminated the DLP Cinema license. But they're coming back as a company, and doing really well. We're talking to others; we may end up looking at a third partner at some point in the future.

What do you say to regular DLP partners who market their projectors as being good enough for cinema-grade projection?

I say that the DLP Cinema solution is the only one that has been endorsed by Hollywood for the showing of first-run movies in movie theaters. There's a lot in that product that was built into the solution during five years of working with the industry.

We recently announced the latest version of DLP Cinema technology, which is called "m15." It's in production and available in DLP Cinema projectors from Barco and Christie. These enhancements include a color-management system called CinePalette; it's an expansion of the color palette of the video projector to emulate film systems. There is no other digital projector that does that today. So when you show the other projectors to cinematographers, they're not seeing all of those same colors.

I've seen those demonstrations where someone's brought in a regular video projector and said 'Look, I can fill a 30- or 40-foot screen with an image, why isn't this good enough for digital cinema?'

Another m15 enhancement is called CineBlack Contrast Management. It's a higher-contrast ratio, extended bit-depth modulation technique for smoother tonal scale between dark blacks and gray shades, and even up to whites. So there's these very subtle things that the golden eyes recognize that we've built into the DLP Cinema product.

We have another technique called CineCanvas Image Management, which provides for text-generation for subtitling, and that's an important aspect that really hasn't been shown off yet for digital cinema projectors. The goal is to have the subtitling rendered in the projector. So you just send a graphics file along with the movie and whatever the language track happens to be, the subtitles will be rendered on the fly so that you can have a much simpler way to get subtitles to international locations.

And then we developed a local-link encryption technique that we've been working with the server companies on called CineLink Security Management. It's a local encryption technique that could take on some other kinds of security features in the future, but it's a way to secure the content coming out of the server into the projector.

Bottom line, DLP Cinema has the features that Hollywood and the exhibitors require.

Right. There's a lot of technology in today's DLP Cinema projector. In fact, we're still bringing those features out through software upgrades, which will enhance hardware solutions far into the future as the digital cinema infrastructure builds out to support them. In fact, a lot of the things we're talking about here—certainly in the subtitling domain and even in the color processing—extend all the way into postproduction. So there's work that has to be done there to take advantage of those features.

Can you elaborate on the postproduction aspect?

One area there is metadata technique: How do you deliver metadata that can carry subtitling or carry color information? How do you embed that into the data stream, or how do you get it on the digital print to the projector or server in the theater? There's not a well-defined technique for that today.

There are SMPTE committees debating various techniques for metadata. We're monitoring that and are working within that activity to get a solution in the not-too-distant future.

What about standards? Digital Cinema Initiatives, the consortium of seven studios, has raised concerns about standards, saying that the evolution of digital cinema has been "manufacturer-driven" but not "quality-driven." How do you respond to such concerns, including those of the National Association of Theater Owners [NATO]?

I think NATO is concerned about how digital cinema is moved forward. Are studios going to help make this transition happen? Is equipment going to be designed such that it's not obsolete in three years? Is there going to be standardization?

But if you really look at what the manufacturers are doing today, they've created a very interoperable system that's defining a broad array of standards that are extensible well into the future with a very flexible, programmable solution that's proving to be very reliable. The interesting thing is that the studios have been intimately involved with this activity and manufacturers have been guided by their requirements. It's going to take time for everyone to realize that and really understand the capabilities that are being deployed right now. You've got the best companies in the world working on some of this, and we're seeing a lot of great engineering going into this. And I think that as people look at it, and as NATO becomes more aware of what we're doing, we're going to see some of their stronger positions softening.

How do you respond to concerns that digital projection is too expensive?

We're at the very beginning of this market, and what we're seeing are the early products that show the capability. Manufacturers will invest in cost-reducing these products when they see customers emerge. The thing that's delaying customers is the expectation that studios would subsidize in some capacity the deployment of digital cinema systems because they are the ones that—at some point in the future—will realize a cost savings.

Until there is a business case worked out, which the studio joint venture presumably could have some impact on, we probably won't see a broad commitment to a digital cinema deployment unless some of the fundamental business dynamics change in some way (Figure 15-2).

For example, with *Star Wars II*, there's the potential of incremental ticket receipts if a significantly greater number of people go to digital screens instead of film screens. Another way for exhibitors to recoup the potential capital expenditure in digital cinema is alternative content—HD sports, Broadway shows, concerts.

I think we'll continue to see some sort of modest roll-out for the time being. Exhibitors are saying, "If there's a chance that studios might pay the cost of conversion to digital, then I don't want to go too far in my deployment plans." It's a very competitive business. Other exhibitors may hit on certain business models that allow them to justify moving more quickly.

And as we see those things happen, manufacturers will be able to make targeted cost-reduction investments because a market will be forming. And that will drive costs down. It's an equation that's balanced on both sides; as the market opportunity grows, there's a potential for more conversion. The investment will grow to enable cost reduction.

Many companies have proposed business models for digital cinema conversion; who will win?

Who knows? What's certain, though, is that there's far more attempts at creating a business model now than there ever has been in the past. And some of those are going to work, and some of those will probably have to be modified. That's what competitive markets are all about.

We're seeing a fundamental change in the theatrical exhibition and distribution business that's a result of a new technology. It's been a long time since that business had to deal with such major change, so it will take time for that industry to really be able to deal with this transition. You're going to see some players that realize the potential and move aggressively, and hit on the right business models. You're going to see some that sit on the sidelines and wait. But that's just like any start-up.

It sounds like you're encouraging the motion-picture industry to take some chances.

If you're scared of digital cinema, you're probably scared of change, and that's not just digital cinema. The exciting thing to me is that there are enough people who are embracing change and saying, 'I don't have all the answers to this, but I'm going to understand it, I'm going to see what it means to my business, and I'm going to get involved.'

We can see it by the exhibitors that got involved with this two years ago, and we can see it by the breadth of exhibitors that are involved with *Star Wars II*. All forms of entertainment media are, one way or another, going digital. The consumer views that as a benefit. Opportunity is born in change. Some people are going to capitalize on that and some won't. So you have to decide on where you want to play.

The Future of the Movies

When *Star Wars Episode I: The Phantom Menace* had its historic digital screening on June 18, 1999, only four commercial movie theaters in the United States were outfitted with digital projectors—and those were temporary installations. Today, there are close to 100 permanent digital cinema screens in the United States.

That may not seem like very many, given the more than 36,000 screens in the United States as of 2001. (The Motion Picture Association of America counts *screens* because most theaters have more than one.) Nevertheless, the transition from the century-old standard of 35mm film projection to the new world of cinema servers and digital projection is slowly gaining momentum in movie theaters around the world. And on college campuses, in coffee houses, and across rural regions of countries such as China and India, the availability of low-cost, high-light-output digital projectors (originally designed for corporate users) are being combined with consumer DVD players to create a growing number of entrepreneurial indoor and outdoor cinemas.

In mainstream movie houses, however, the basic, underlying technology of motion-picture exhibition has gone pretty much unchanged for nearly a century. A long, transparent strip of plastic with sequential photographic images on it is fed from a metal reel, past an intense, shuttered light source, and wound back on to a take-up reel. A projection lens directs the light from these moving images onto a screen at some distance, enlarging the original film frames to many times their original size.

Color, widescreen pictures, digital multitrack sound, and many other innovations have been added over the years, but today's movie projectors would still be recognizable to Thomas Edison. This is a remarkable fact when one considers the profound changes seen in other entertainment technologies during the same 100 years (Victrolas to iPods, crystal radios to satellite radio, stereopticons to DVDs, etc.). But ironically, theatrical motion pictures—which reign supreme in the hierarchy of moving images in terms of picture information and emotional impact—still employ the same basic technology. Perhaps the digital cinema revolution is, in truth, long overdue.

The Digital Difference

Projecting movies digitally has numerous advantages over film projection. Film prints wear out quickly, degrading the beauty of the original image the filmmaker intended. Film prints are expensive (costing at least $1,000 each). They must be shipped to theaters and then sent back to the distributor to be destroyed. They are easy to pirate. And the chemicals used in processing film are potentially harmful to the environment.

When motion pictures are projected digitally, however, they exist as nothing more than a stream of data fed from a server, which is in fact an array of hard drives upon which the movie has been stored in digitally compressed form. This compression is necessary to fit all of the image data comprising a full-length feature film (see Chapter 4). Cinematographers and projector manufacturers may argue over whether this compression negatively impacts the original picture quality of movies, but evidence indicates that audiences prefer digital projection. With no actual strip of film passing through a projector to be scratched and dirtied, the movie never degrades. Colors don't fade, images don't jitter, and the audio is delivered in uncompressed multitrack perfection. Audiences watching the 100th screening of a digitally projected film see the same pristine quality as the audience at the film's first screening. And the original beauty of the film's images—regardless of whether it was photographed on a film format, a digital HD videotape format, or some other digital cinematography medium—is presented with much greater fidelity.

According to Texas Instruments (TI), manufacturer of the DLP Cinema "black chip" Digital Micromirror Device that generates the images in nearly every commercial digital motion picture projector (see Chapter 15), their digital cinema technology has been used to project more than 150,000 screenings to more than 14 million people worldwide since 1999. TI's research indicates that 85 percent of viewers described the image quality they experienced as "excellent," and no fewer than 80 percent of audiences stating that—given the choice—they would prefer to see movies projected digitally, as opposed to from film.

A 2003 *Screen Digest Digital Cinema Audience and User Preference Study*, meanwhile, states that research done in North America and Europe also indicates that audiences prefer digital cinema projection over traditional 35mm film projection. This is particularly the case with films containing a high proportion of digital imagery, such as *Spider-Man* or *Van Helsing*. It's also true in the case of films that were photographed entirely in digital, such as *Star Wars Episode II* and *Episode III*, and with films that were digitally generated using computer-animation technologies, such as Pixar's *Toy Story*, *Finding Nemo*, and *The Incredibles*.

Increased Efficiency

Picture quality isn't the only advantage of digital cinema servers and digital cinema projectors. When a film exists solely as a body of stored digital data (as opposed to reels of 35mm film), it can be encrypted in multiple ways, which makes piracy far more difficult. And while most piracy consists of audience members shooting movie screens with camcorders, digital cinema offers a means of thwarting that as well. Digital watermarking, developed by several research groups, including the Princeton, New Jersey-based Sarnoff Corp., enables movie studios to reliably trace camcorder-pirated films back to the theaters they were recorded in, even after the kind of low-bit-rate compression these films are subject to when distributed on the Internet. Digital watermarking codes—invisible to the human eye but not to computer analysis—are inserted in the digitally projected motion picture. And when detected in pirated DVDs, tapes, or on the Internet, authorities can tell when and in what theater the movie was videotaped from the screen. This makes exhibitors far more accountable for policing their theaters.

Motion-picture distribution and delivery is also being revolutionized by new digital technologies. And instead of having to be physically transported to theaters in large, heavy film cans by fleets of trucks, digital movies can be distributed to theaters via satellite. This technique was pioneered by Boeing Digital Cinema in 2000; Access Integrated Technologies purchased the assets of Boeing Digital Cinema in March 2004, and continues to distribute movies this way. Other techniques for digital distribution include fiber optic networks and physical media such as DVD-ROM discs, single computer hard drives, and servers. Needless to say, satellite and fiber are the most economical form of motion-picture distribution, because they eliminate the need for couriers. Estimates vary, but one study (Screen Digest's 2000 *Electronic Cinema: The Big Screen Goes Digital*) calculates that the current worldwide cost involved in distributing film prints is $5 billion; digital motion-picture delivery has the potential to reduce that figure by 90 percent.

Other benefits of digital motion-picture distribution and playback include the ability for multiscreen theaters to add or delete the number of screens showing a particular title at the click of a mouse, depending on audience demand. Instead of needing multiple prints to show a film in multiple cineplex or megaplex auditoriums, the playback stream from a single server can be routed to additional screens simultaneously, depending on which movie is selling the most tickets. Film trailers can be easily updated, with new coming-attraction previews arriving via satellite, overnighted hard drive or DVD-ROM, or even via the Internet. And, using a simple PC interface, this content can be inserted into the server in whatever playback order the exhibitor desires—rearranged at will with the ease of moving blocks of text around in a word-processing program.

Cinema servers can also be programmed to record the exact dates and number of times that movies are shown, and can then transmit that data back to the studio or content-rights holder for an accurate accounting of what they're owed from the exhibitor. And when the contracted exhibition period ends, the server can be programmed to automatically erase the film so it can't be played again.

Servers—which are basically PCs with ample storage—interface well with theater-automation systems (Figure 16-1). The entire process by which a movie is shown can be managed very precisely when a theater is fully digital. One central PC can dim house lights on schedule, open the curtains, start the projector, and even turn on the popcorn machine. Theater employees can then concentrate on selling tickets and candy.

Given all these advantages, it's not surprising that no less a force in film than Kodak has announced its support of digital cinema. Movie exhibitors attending the annual ShowEast conference in Orlando in October 2004 learned about the Kodak Digital Cinema Solution for Feature Presentations. At the core of this system is the Kodak CineServer, which is designed to enable exhibitors to decrypt, decompress, and send motion-picture content at up to 2K (2,000-line) resolution to digital projec-

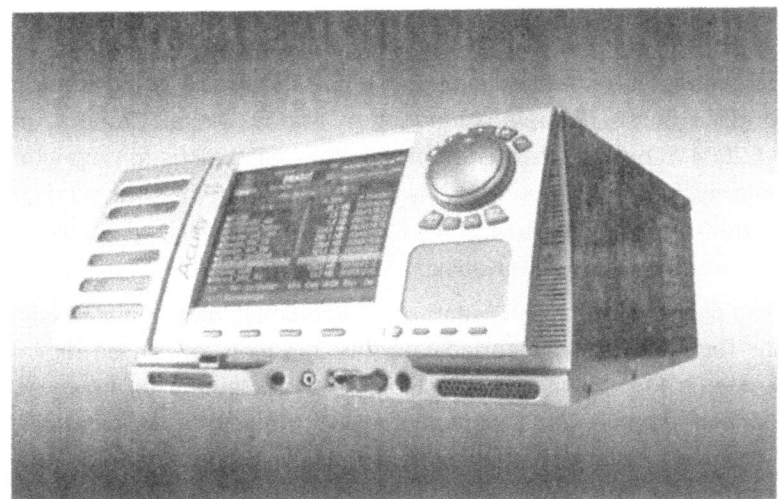

Figure 16-1

The QuVIS Cinema Series servers replaces traditional reels of 35mm film.

tors. And Kodak has indicated its commitment to a 4K system, which in the opinion of many in Hollywood is the equivalent to the picture information in 35mm film. The Kodak system will deliver movie content to theaters on DVD-ROMs encoded with Digital Source Master files created by LaserPacific, a major Hollywood HD postproduction, compression, and mastering facility Kodak purchased in 2003 (Figure 16-2). Clearly, the world's number one manufacturer of film has its sights set on a digital future.

Figure 16-2

Kodak's Digital Cinema Pre-Show Systems enable exhibitors to customize cinema content for specific screens, ratings, dates, and films via a PC interface. The system also provides exhibitors with automatic verification each time a movie plays.

"This is a significant step in our continuing commitment to help assure the evolution of digital cinema in ways that best serve the interests of exhibitors, studios, the creative community, and the movie going public," explained Bill Doeren, General Manager of Kodak Digital Cinema, at ShowEast 2004. "Our approach is to offer a menu of choices to help simplify complexity, to provide the solutions they [exhibitors] need to 'go digital' when they're ready, and to assure them of the quality and attention they expect when working with Kodak."

Leading film laboratory, DVD replicator, and digital intermediate facilities owner Technicolor has also indicated its intention to provide digital cinema systems for exhibitors. Technicolor Digital Cinema is, however, delaying its plans until Digital Cinema Initiatives (the seven-studio consortium determining technology specifications for future distribution and projection of theatrical films) finishes its studies.

Microsoft as well has staked its claim on digital cinema, announcing in April of 2003 that the entire Landmark Theater chain (53 theaters with 177 screens in 13 states) would be permanently outfitted with PC-based cinema systems for playback of movies based on the software giant's Windows Media 9 Series digital media format.

Given such high-level endorsements of digital cinema, one might expect exhibitors to be eager to junk their 35mm film projectors and install the latest cinema servers and digital projectors. But this is not happening because current digital cinema technology is still more expensive than its reliable, familiar "celluloid" counterpart. And, more important, the question remains as to who, exactly, will pay for this transition.

Why Fix What Isn't Broken?

The movie business has been a profitable one throughout most of its history. Despite exhibitors overbuilding megaplexes in recent years—which plunged several of them into bankruptcy—the movie business is generally profitable. A series of recent mergers and acquisitions, meanwhile, has restored a large measure of stability to the exhibitor business.

The fact that film-projection technology has remained relatively unchanged for many years is based on a simple fact: It *works*. Film is a mature, effective, and relatively simple technology. Modifications have been developed over the years as needed: large film platters to eliminate reel changes, interfaces with house-automation systems, digital sound formats, etc. But—as noted previously—there hasn't otherwise been a great need to fundamentally change the theatrical film-projection paradigm. And in the opinion of many—although not all—exhibitors, this still holds true, digital cinema or not.

Film emulsions, meanwhile, have never been more advanced, yielding detail and color that provide a palette for artistic expression of unprecedented variety. Digital multitrack sound (see Chapter 8) further heightens the experience, occasionally to the point of assaulting the eardrums. A motion-picture projector costs on average $50,000 and lasts 25 years. The century-old 35mm film format is the world's most successful imaging standard; take a reel of 35mm film to any commercial movie theater on earth and the projection system there will probably be able to display it.

A theatrical digital cinema projector, however, costs $150,000 and works with a server that could be made obsolete at any time by the ever-advancing computer technologies described in Moore's Law. Digital projection itself is a moving target; as noted above, many in Hollywood feel that anything less than a projector capable of displaying the equivalent of 4,000 lines of resolution ("4K") is unacceptable (see Chapter 5). Texas Instruments, JVC, and Sony all have excellent projection technologies, but all are *incompatible* with one another. What theater owner in their right mind would want to *replace* the reliability of 35mm film for a very new technology that could be orphaned in just a few years by rapid technology advances or a lack of marketplace acceptance? And—again—who will pay for this conversion to digital cinema?

The answers are: *Not that many theater owners* (as of right now) and *Nobody knows*. Paradoxically, however, many of those same exhibitors are installing digital projectors and servers anyway in their projection booths. These installations are, however, not intended to replace 35mm projectors, but to go alongside them. Why? The reason has little to do with feature film exhibition and everything to do with *advertising revenue*.

Regal Treatment

European cinema audiences have long been used to seeing beautifully produced advertisements on their theater screens prior to feature films. This has not traditionally been the case in the United States, although companies such as National Cinema Networks have developed systems for Internet delivery of quality in-theater digital advertising in recent years. More typical of in-cinema advertising in the United States, however, has been content projected in low quality by small and relatively dim corporate video projectors or even 35mm carousel slide projectors. A new generation of affordable—yet more powerful—digital projectors, however, have the potential to change that.

Although high-end projection based on TI's DLP Cinema technology is deemed essential for theatrical feature-film presentation by studios and exhibitors, that's not necessarily true of in-cinema advertisements. Lower-resolution technologies such as TI's non-cinema-grade DLP devices in smaller, less expensive projectors (costing tens of thousands of dollars, as opposed to more than $100,000) are still capable of presenting a far better image on screen than what has been used in the past. And when combined with a PC linked via the Internet to a central network server, new trailers and ads can be automatically updated and sent to theaters and customized for a particular film's audience. This lower-cost form of digital cinema offers the promise of new advertising revenues from national companies eager to reach mass audiences that haven't been sold to before very effectively.

Regal CineMedia Corp., the media subsidiary of Regal Entertainment Group (the largest movie exhibitor in the United States), awarded Christie Digital Systems—a leading manufacturer of DLP and DLP Cinema-powered projectors—a contract in late 2002 to supply digital projectors and displays for the world's largest theatrical networked digital projector system. The contract includes Christie digital projectors, a networking system known as ChristieNET, and in-lobby plasma (flat panel) screens. The entire project was deployed as part of Regal CineMedia's national Digital Content Network (DCN), which was installed in nearly 400 Regal movie theaters in 19 of the top 20 markets, and 43 of the top 52 markets. Christie RoadRunner L6 DLP-based projectors are used to display short-form entertainment, sports, education, and other types of preshow programming, as well as local and national ads. Marketing and advertising content is also shown on plasma screens in theater lobbies.

Regal's digital in-theater content is branded as *The 2wenty*, which consists of a 20-minute ad segment shown before the movie's advertised start time. These long-form ads are designed to be entertaining to blunt U.S. movie audiences' dislike of "TV commercials" in theaters. *The 2wenty* features select original programming content from TBS, NBC, and other sources. It is digitally distributed by a sophisticated Network Operations

Center to approximately 4,500 theater screens and more than 1,400 42-inch high-rez plasma displays in theater lobbies. On January 17, 2003—approximately one month after Regal first announced its plan—Christie formed a Cinema Systems Group, a new business unit to administer turn-key solutions of the kind included in the DCN deal.

The Regal CineMedia digital network currently delivering *The 2wenty* to nearly 400 theaters and 4,700 screens is reported to have earned approximately $74 million in 2003, an increase of more than 236 percent in Regal's advertising revenue over the previous year. And Regal is not alone.

Starting in November 2003, a company called Cinema Screen Media (CSM) began purchases of Kodak's Digital Cinema Pre-Show systems, which Kodak is installing in Harkins Theaters and other chains. As the book goes to press, CSM has purchased more than 1,300 such systems. These systems—not to be confused with Kodak's Digital Cinema Solution for Feature Presentations—will deliver CSM-produced/Kodak-prepared pre-show content to more than 1.5 million Harkins moviegoers each month. The content is described as a mix of marketing messages, entertainment, and alternative programming. And it is very possibly in this last category that digital cinema's most promising potential lies.

Alternative Programming

Movie theaters are peculiar places. They sit empty for most of the day, and usually attract patrons only during evenings and weekends. But imagine if there were new and compelling reasons for new kinds of audiences to patronize theaters. This is the potential of alternative programming. After all, why should feature films be the only content shown in movie theaters?

A great advantage that digital cinema projection systems have over film projectors is their ability to show more than just movies. This digital content can include live or recorded HD of sporting events, rock concerts, operas, or any other program material people enjoy seeing on the larger-than-life dimensions of commercial cinema screens. Numerous successfully presented alternative programming demonstrations in recent years have included Broadway musicals (by Broadway Television Networks), World Cup Soccer (by Brazil's TV Globo), and a David Bowie rock concert (by satellite services provider Kingston inmedia and Tandberg Television). Regal Cinemas presented satellite broadcasts of five concerts in 2003 from such groups as The Grateful Dead and Coldplay, as well as sporting events. Other potential forms of alternative digital cinema programming include multisite corporate meetings, political rallies, educational events, and religious services. The idea is not a new one; live satellite video of boxing matches projected in special venues dates back at least as far as the 1970s—and that was with far less picture quality than is possible with today's digital cinema systems.

European movie industry experts tend to refer to cinematic preshow content, ads, and alternative programming as "E" (electronic) cinema to differentiate it from "D" cinema, or the digital projection of feature films. In the United States, John Fithian, President of the National Association of Theater Owners, referred to this form of content with his own term, *Other Digital Stuff*, during a speech he gave at the 2003 Digital Cinema Summit at the National Association of Broadcasters Convention, in Las Vegas. Fithian insists that the abbreviation for his term, "ODS," should not be pronounced as *odious*. Some digital cinema proponents have, however, argued that anything other than the presentation of feature films in movie theaters is indeed undesirable, potentially threatening to the longstanding and special nature of the cinema experience.

Nevertheless, alternative programming is a potentially enormous area of the digital cinema revolution that has only just begun to be explored. Whether audiences will pay a premium to sit in comfortable stadium-style seating—complete with cup holders—in today's megaplexes to watch cinema-size, digitally projected live HD of major football games or other special events remains to be seen, for now.

The End of the Beginning

Some media pundits have predicted that the proliferation of home theaters, digital TV broadcasting, high-definition widescreen televisions, and other new home-entertainment appliances (such as HD DVDs) will spell the death of cinema. Audiences, they say, will opt to enjoy the comfort of their homes and forego the hassle and expense of babysitters, driving, parking, tickets, overpriced snacks, dirty theaters, uncomfortable seats, and inconsiderate cell-phone-using movie audiences.

Given the wretched state of some movie theaters, there is no doubt that the theatrical movie-going experience lacks allure for many people. There are also many plush new megaplexes with comfortable seating that make moviegoing more enjoyable than ever before. Statistics on theater attendances don't indicate a massive decline. If anything, the reverse is true. Also true is the fact that the movies offer a much-needed social outlet not to be found in many other places. Movies provide a convenient excuse for dating and courtship, a shared entertainment event similar to traditional theater, and a dramatic, larger-than-life visual presentation not generally available elsewhere. The dimensions of the average motion-picture theater screen (approximately 30 feet across) tend to fill most of the human visual field, heightening our suspension of disbelief and providing a compelling entertainment experience.

The digital cinema revolution is still in its infancy, its full effect yet to be determined. But whether its greatest impact will come in the form of grass-roots moviemaking and ad hoc theaters set up in faraway locales, college campuses, or local community centers, or in Hollywood extravaganzas delivered through virtual reality goggles that immerse audiences directly into the action, humanity's long affinity for storytelling and visual communication suggests that digital cinema's future will be bright indeed.

Epilogue

The Age of Digital Cinema has only just begun, with new developments and technologies being announced at an increasing rate. There is a tremendous thirst for objective information on this topic across a wide sector of the entertainment industry, from the highest echelons of Hollywood to the most basic indie filmmakers. Everyone is being challenged to stay informed of this historic and ongoing integration of digital technology into the motion-picture industry.

In answer to this need, the Digital Cinema Society—a nonprofit corporation dedicated to educating and informing all interested parties about digital motion picture production, post, delivery, and exhibition—was created in April 2004.

The Digital Cinema Society is custom-made to address the information needs of producers, directors, cinematographers, postproduction specialists, exhibitors, indie filmmakers, and everyone else with a professional interest in motion pictures. This nonprofit organization is open for membership to all of the above segments; its purpose is to objectively examine all media, solutions, services, and technologies without favoring any one brand, service, or format over another. There are many definitions of digital cinema, and all of them are explored by the Digital Cinema Society.

The Digital Cinema Society welcomes your input. Cofounders James Mathers and Brian McKernan invite you to visit www.DigitalCinemaSociety.org/join for more details about how you can participate.

Index

NOTE: Boldface numbers indicate illustrations.

ABOUT THE AUTHOR

Brian McKernan has been writing about media technologies for more than 25 years, and is the founding editor of *Digital Cinema* magazine. The coeditor of McGraw-Hill's *Creating Digital Content*, he was editor of *Videography* magazine from 1988 to 1999 and the *Age of Videography* twentieth anniversary edition (1996). He originated Day Two of the Digital Cinema Summit at the annual National Association of Broadcasters' conference and is the cofounder of the Digital Cinema Society.